A.H. Haley was born in Anglesey in 1912. In 1932 he qualified as a teacher. He enlisted in the RASC in 1933 and transferred to the Education Corps two years later. From October 1937 to January 1943 he was stationed with the Garrison in Mauritius where he experienced some of the effects of long service on an outstation. His interest in the Crawley Affair started in 1967 when he came across an account of the Mhow court martial in the 1863 volume of the CORNHILL MAGAZINE. With his own experience of long service abroad he 'at once became hooked and lived with the Crawley Affair for the next three years.'

The Crawley Affair – A.H. Haley

India in 1861 was not a place where the whims of a British Cavalry Colonel could be lightly ignored. When Colonel Crawley took over command of the 6th Inniskilling Dragoons at Ahmednugger in April of that year the Regiment, already at odds with higher authority, was subjected to a man with a remarkably volatile temper and acid tongue. In a very short time the officers' mess was split into two opposing camps – a majority burning with resentment against their Commanding Officer and a few who supported him.

Thus began a chain of events which led to the death of the Regimental Sergeant Major and which ended at a Court Martial in Aldershot on which was focussed the full glare of Victorian publicity, at once prudish and prurient.

Mr Haley skilfully weaves together the personal tragedy of RSM Lilley and his wife with the invidious position of Senior Officers who became personally involved in the problems of Colonel Crawley and were gradually manoeuvred into supporting him in order to maintain their own reputations.

This is a remarkable and compelling story which has all the elements of classical tragedy. Under Mr. Haley's dexterous handling, one becomes hypnotised by the inexorable approach of one disaster after another, and longs to invervene and ward off the invevitable. This is indeed history brought to life.

The Crawley Affair

The Crawley Affair

A. H. HALEY

BULLFINCH PUBLICATIONS

Published By
BULLFINCH PUBLICATIONS,
245 Hunts Cross Avenue,
Woolton, Liverpool 25.

Produced By
THOMAS LOUGHLIN,
Mulberry House, Canning Place,
Liverpool L1 8JG.
Tel: 051-709 0818/9

Cover designed by Peter Charles

First Published 1972 Seeley Service & Co. Ltd London

This edition 1988 Bullfinch Publications

British Library Cataloguing in Publication Data

Haley, A.H. (Arthur Henry). *1912 –*
The Crawley Affair.
    1. Great Britain Army. Thomas Crawley, 1818-1879
    I. Title

355.3'32'0924

ISBN 0 9511427 3 9

Contents

THE ALDERSHOT COURT-MARTIAL

Acknowledgements

In the writing of this book, I have drawn mainly upon the following sources:

Public Record Office.
  W.O.33/13 and W.O.33/14.
  Records of officers.

Sessional Papers of the House of Commons.
  1863. Volume 33. Pages 1–199.
  1864. Volume 35. Pages 317–465.

Somerset House.

Hansard.

Dictionary of National Biography.

Various Regimental Histories and military books. In particular *Records of the Inniskilling Dragoons*, by E. S. Jackson.

Newspapers and Magazines:
  The Times. The *Manchester Guardian*. The *Liverpool Daily Post*. *Grantham Journal*. *Lincolnshire Chronicle*. *Nottingham Review*. *The Spectator*. *Illustrated London News*. *Cornhill Magazine*. *Temple Bar*. *British Quarterly Review*.

My thanks are also due to:

The Ministry of Defence Library for advice and for making books available to me,

Lt-Col G. A. Sheppherd, MBE, the Central Library, the Royal Military Academy, Sandhurst, for T. R. Crawley's record as a cadet,

Major F. H. Robson, Home Headquarters, the 5th Royal Inniskilling Dragoon Guards, for information on officer's records,

Major D. Derham-Reid, Curator of the East Lancashire Regimental Museum, for Thomas Crawley's military record, the Reverend J. H. Jacques for letting me inspect the parish registers in Spilsby Parish Church for details of the Lilley family,

Mrs J. E. Egerton for invaluable help with research, and to the Liverpool Central Libraries, particularly the Picton and Microfilm sections where I spent many hours and received unfailing and cordial assistance.

A. H. HALEY

1

An Officer and a Gentleman

Among the thousands who enlisted in the British Army in the year 1807 to meet the Napoleonic threat was a militiaman named Thomas Crawley. He probably came over from Ireland (or at least had Irish connections), and being a cut above his fellows in strength of character and education, rose in due course to the rank of R.S.M.

A turning point in his career came in July, 1813, when he arrived with Wellington's Grand Army before St Sebastian, the last stronghold to bar its victorious march into France. His Regiment (59th Foot) took no active part in the abortive first assault on the city walls but was given a supporting role in the trenches.

While waiting there Crawley received a severe gunshot wound, the effects of which may be seen either as a tragedy or as the intervention of a kindly fate. He was lamed for life—but was effectively removed from the force that carried out the successful attack on the fortress a month later, when his Battalion lost half its personnel including twenty-one officers.

For the lucky survivors (even the maimed) prospects for promotion were unusually bright and a few days later Crawley was commissioned as adjutant. He remained on the active list until the following May when he was retired on the half-pay of a lieutenant.

Unlike most of the ex-rankers who 'made it', he managed to hang on to his respectability and ambition through the lean years that followed Napoleon's defeat. He had won the right to call himself 'Officer and Gentleman' and evidently decided to use it (as another might use a modest legacy) to found his family's fortunes—that is to set his sons on the military promotion ladder with a better chance of climbing it than he had enjoyed.

His first success came in 1827 when he obtained a commission for his eldest son, Henry, in the 20th Foot as ensign 'without purchase'.[1]* Then in 1832 he entered his second, Thomas Robert (the subject of this narrative) as a gentleman cadet in the Royal Military College, Sandhurst, at the greatly reduced fees payable for the son of a poor but deserving officer.†

The boy bore stoically the rigours of the two-year course and, after scoring high marks in the examination specially set for the purpose was awarded a free commission in the 45th Foot. At sixteen he had already shown the more estimable qualities that characterized him later—a sharp intelligence, dedication to his profession and a capacity for acquiring friends who could be useful later (an invaluable asset for an impecunious officer).

In 1839, not long after his Regiment arrived in India, he obtained a transfer to the 15th Hussars. This was advancement of sorts in that his new rank of cornet, although equivalent to the infantry's ensign was worth an extra £400 at the regulation price.

Three years later came a stroke of luck. A captain in his Regiment died and, according to the rules, the value of his commission died with him. A lieutenant was promoted to fill the vacancy without paying any purchase money and Crawley took the place of the lieutenant on the same terms.

Meanwhile his intelligence and zeal recommended him to his superiors and he was given a staff appointment, the first of several he was to hold during his 14 years in India. He used them, not only to acquire a minute knowledge of Army Administration, but to study at first-hand the mentality and thinking of the top brass. He also made many friends in high places.

Perhaps the most influential of these, General Sir John

---

* Numbered references in the text refer to the Appendix.
† Two more sons were commissioned. Both died young in India. One was killed at Sabraon, spiking a gun he had captured. The other died from the effects of the climate.

Aitcheson, G.O.C., Madras, to whom he was A.D.C. for six years, later described how the young man used to spend his spare time. He was never idle but would travel for miles around Bangalore, surveying and mapping out the countryside, which, as Sir John said, was important training in reconnoitring for the Hussar.

During a six-month's leave he got himself attached to the 60th Rifles to take part in the Second Sikh War, the only active service he was to see in his long military career. He was doubtless fired by ambition though perhaps not entirely. Lack of the means to keep up with the extravagant tastes of his brother officers may have forced him to find some outlet for his energies.

Either way his enthusiasm was bound to impress those who mattered. But in the absence of hard cash no further promotion came his way and he returned to England in 1852 still a junior officer.

Then, in October, 1854, his fortunes dramatically improved— he bought the command of a troop. Though how he managed to raise the large sum (£2,035) can only be surmised. A possible clue lies in the fact that four months later he married a young lady named Catherine Taylor, the only daughter of a colonel who, like Crawley's brother, had been killed in action at Sabraon in 1846.

The ceremony took place at St John's Church, Paddington; the groom being in his 36th year and the bride not yet 21. Such an age gap was not however uncommon at the time when economic necessity made it only prudent for the breadwinner to marry later in life. And as the widowed mother came to live with Crawley, an arrangement may have been arrived at to secure his promotion and provide her with a permanent home.

Early in 1858, the major of his Regiment was promoted to the newly raised 18th Hussars, a fortunate move that gave Crawley his majority without purchase. The following year the command of the 15th Hussars was up for sale and as the senior officer he was allowed to purchase at the regulation price of £1,600.

He had reached the top of the regimental ladder, realizing the ambition set by his father now long since dead. Luck had played its part, yet he could claim much for his own efforts. And now that success had come he might have been expected to sit back and enjoy it. But for some reason he was not content and decided to push his luck a little further.

He had been in command of his old Regiment for only a few months when he heard, possibly through the headquarters of the Black Market that existed in London for arranging such matters, that a lieutenant-colonel in the Inniskilling Dragoons, stationed at Kirkee in India, wanted to exchange into a cavalry regiment in England.

There are several reasons why he may have found the offer tempting. Keeping up the style of a cavalry officer was less expensive in India than at home, and the offer of the exchange may have carried with it a financial inducement, for Colonel FitzWygram, the Inniskilling officer concerned, was a rich man. The desire to return to India, too, could be strong in a man like Crawley who had spent many years there in his youth.

One difficulty had, however, to be surmounted before the exchange could be effected. For though FitzWygram was a lieutenant-colonel, he was not the C.O. of the Inniskillings, having obtained his rank by a regulation then in force which allowed regiments in India to carry two lieutenant-colonels on the strength instead of the usual one. And being the junior and only second in command, he was seemingly in no position to offer the Regiment in exchange for another.

At the same time, Crawley was hardly likely to relinquish the command of the 15th Hussars and journey half across the world to take up an inferior appointment in a strange regiment. Fortunately, or so it seemed at the time, there was a way out. The C.O. of the Inniskillings, Colonel Shute, had also had enough of India and wanted to come home, though his aim was to sell out and retire on half-pay.

Both he and FitzWygram had spent many years with their regiment, and had the latter wanted to stay in India, its command would probably have been his for the asking. As it was, the next logical contender for the appointment was Major Hunt, another long-serving Inniskilling, who in due course sent in his application.

Briefly then, the situation was this. Crawley and FitzWygram wanted to exchange but only on equal terms. Shute wanted to retire with as much money as he could get for his rank together with the command of the Inniskillings, while Hunt wanted to purchase the command.

The outcome would be very much up to Crawley, the only one of the four near enough to influence the final decision at the Horse Guards. This was how he set about it.

On 24 August, 1860, he sent in his application for the exchange. (FitzWygram's had been dispatched from India some time before.) The following day he wrote from Newbridge in Ireland, where he was stationed, to General Sir John Aitcheson, his old chief at Madras, now retired in London.

My Dear Sir John,

I have but a minute to say that I sent in my papers for the exchange yesterday and that they will probably be in General Forster's hands before you receive this. As it is of the *utmost* importance that I should get the start of Major Hunt in order to get the command of the Regiment if possible, I hope you will kindly excuse my asking you to see General Forster to have the exchange take precedence of the sale of Colonel Shute's Lieutenant-Colonelcy. Pardon the hurry I write in and with kindest regards—

Ever yours—most warmly (scrawl),

T. R. Crawley.

On receiving this appeal from his protégé, Sir John wrote immediately to General Forster, the Military Secretary to the Duke of Cambridge, at the Horse Guards.

My Dear Forster,

Lieut. Col. Crawley is I consider one of the best officers in the Army, and as I speak with many years' knowledge of him, I hope you will do *me* the honour to help him if you can. Major Hunt is his junior by 13 years.

His letter was dated 27 August. On the 29th General Forster endorsed the papers of exchange, 'Has your Royal Highness any objection?' The Duke had none. And so the commands of the two famous regiments were settled to the satisfaction of Crawley and FitzWygram, though it may be assumed that the other two officers involved were in some degree disappointed.

Five months later Shute sold his Lieutenant-Colonelcy to Hunt and retired on half-pay, though only briefly as will be seen later. Hunt held his new rank for a year, all of which he spent in England away from his Regiment. Then he too retired from the Army and disappears from the story.

2

'Impatience of reproof—unaccompanied by amendment'

When Lieutenant-Colonel Charles Cameron Shute was given the command of the 6th Inniskilling Dragoons early in 1858, the appointment must have appeared sound and logical.

He had been with the Regiment for the whole of his military career and second in command for the last four years. In the Crimea he had been recommended for the V.C. (though not awarded it) for his part in that remarkable prelude to the Charge of the Light Brigade when the Heavies, spearheaded by the Inniskillings and the Scots Greys, routed the mass of Russian Cavalry. During the latter part of the war he held the post of Adjutant-General to the Cavalry Division and throughout 'he never was absent from duty'.

As a cavalry officer he was generally thought to have few equals. All the Inspection Reports of the period following the return of the Regiment from the Crimea in 1856 bear glowing tributes to his ability, and the Inniskillings under his influence were also highly praised for their all-round efficiency. The following encomium from the Earl of Cardigan in July, 1858, just before they sailed for India, was later much quoted and may be regarded as typical.

'This Regiment is a model and a pattern for the Cavalry of the service, no other regiment now in the United Kingdom being able to move with the same celerity, precision and correctness of drill united together.'

Not of course that these official assessments should be
accepted entirely without reservation. For within the circle of
high-ranking officers it was not usually considered necessary to
turn the light of criticism too strongly on ones own kind. An
inspecting general, well disposed to a commanding officer, might
be expected to refrain from inquiring too closely into some unfavourable aspect of the regiment and would avoid any official
mention of it altogether. Lord Cardigan, for instance, in his
report, vouched for 'the unanimity and good understanding'
amongst the officers of the Inniskillings, a statement hard to
reconcile with an occurrence of only a few weeks earlier.

This involved a troop captain named Weir whose career was
on all counts remarkable. Commissioned from R.S.M. he had
attained his present rank in the short space of six years and in
the process had paid out no purchase money. Such rapid promotion of an ex-ranker in an exclusive cavalry regiment was exceptional enough; it was all the more so in that Weir's advancement to cornet, lieutenant, captain—had all taken place in
England and not when the Regiment was abroad or to fill vacan-
cies caused by the usual exodus of wealthy officers when a
foreign posting became imminent.

He had been promoted on the recommendation of Shute's two
immediate predecessors so it may be assumed that both had a
high opinion of his abilities. They must also have considered
that he had managed to adapt to the enormous social change
from the Sergeants' to the Officers' Mess, so often the stumbling
block to the man who earned his commission from the ranks.

Colonel Shute too must have known Weir well, for they had
served together for many years. It is therefore surprising to find
that only two months after taking over the Regiment he adopted
the extreme course of referring the ex-ranker to the C.-in-C.,
the Duke of Cambridge,[2] on a matter of discipline.

'Captain Weir,' he wrote, 'has too well succeeded in sowing
a vast amount of dissention amongst the staff of the regiment to
which he owes everything, and in which, from being an indifferent R.S.M. and a very moderate adjutant, he has risen from the
ranks to be, in only six years, a captain without purchase.'

The Duke, understandably averse to being drawn into such
petty controversies, replied that he would have no hesitation in

removing Captain Weir and any other officers who persisted in disrupting the Regiment by their quarrelling. At the same time he expressed surprise that Colonel Shute had not settled a disciplinary matter of that sort himself. As far as he was concerned, however, the matter was closed and he seems to have maintained his high opinion of Shute as a commanding-officer.

But as this was by no means the last to be heard of the letter, it may be noted that the sentiments expressed in it could be said to imply a good deal more than the condemnation of a trouble-maker, namely that the writer was incapable of controlling his officers; that he was against promotion on merit; and that his predecessors who recommended Weir for promotion were ignorant of the qualities required in a N.C.O., let alone an officer.

That Shute in his role of 'new broom' was merely trying to get rid of an officer he disliked, may or may not be accepted in the light of future events. If indeed that was his object, the sequel to his failure was ironic. Sailing orders for India came at the end of the summer and with the inevitable rush of wealthy officers out of the Regiment to avoid the rigours of a foreign tour the ex-ranker was rocketed from near the foot, right to the top of the list of troop captains.

The Regiment arrived at Bombay in October, and immediately travelled the few miles to Kirkee where they began the essential task of breaking in 'remounts'. But before this could be completed, the Mutiny, the reason for their hurried departure from England, was virtually over and the inevitable sense of anti-climax set in. After the general regret at being deprived of action had subsided, however, the Regiment settled down happily enough to cantonment life, the officers particularly finding relief from boredom in the social amenities of Poona which was within easy reach.

As with all units in India, in semi-isolation, surrounded by an alien culture, some deterioration in morale was bound to occur, though the process was well understood by visiting generals who were usually prepared to make allowances accordingly. But there were at this time certain factors quite out of the ordinary.

Following the Mutiny, important changes in the organization of the Army, brought about by the greater responsibility assumed by the British Government for the running of Indian affairs, were being made, the foremost of which entailed the amalgamation of the Queen's Forces with those of the East India Company; and at this moment in their history it was the fate of the Inniskillings to come under the immediate surveillance of Sir Hugh Rose,[3] one of the heroes of the recent fighting, who was soon to be entrusted with carrying out this difficult task.

Sir Hugh first inspected them as their area commander in March 1859 after they had been at Kirkee about eighteen months. His report praised highly their efficiency and smartness which spoke well of the 'professional qualifications and untiring zeal of Colonel Shute', adding that *Colonel Shute makes good officers*.

He made one criticism. There had been far too many courts-martial and as ten had been on N.C.O.s, he suggested that not enough care had been taken in their selection.

All the officers gave their Commanding Officer proper support with one exception, 'who, from frequent sickness, the result of his own intemperance, has been almost useless'.

Like all previous reports on the Regiment when commanded by Shute it was a good one, though obviously written by one who regarded inspections as no mere formality and was ready to make official criticisms without pulling his punches.

In succeeding months, however, Sir Hugh Rose's opinion of the C.O. and the officers of the 6th Dragoons underwent a radical change, the reasons for which appear trivial enough, but as they lie at the root of the larger conflicts in which the Regiment eventually became involved, require some description.

The posting to India had, as stated earlier, resulted in a large turn-over of officers, some of the newcomers being of a lower social standing than those they replaced. A few exchanges had also taken place since it arrived.

At one extreme of the social scale were those with the traditional background of cavalry officers including Colonel Shute, who had been to Winchester, Lt-Col FitzWygram to Eton, and the hon Captain Charles Wemyss Thesiger, second son of Lord Chelmsford, who had also been to Eton. At the other were the ex-rankers—Captain Weir, Captain Anderson at one time

2

Quartermaster in the 13th Hussars, who joined the Inniskillings as Riding Master and was now in charge of a troop, and Cornet Robert Davies, an ex-sergeant in the 11th Hussars (Cardigan's Regiment), who had been wounded in the Charge at Balaclava. The two appointments of Riding Master and Quartermaster normally reserved for ex-rankers were held respectively by Joseph Malone and Charles Wooden.[4] Each had been awarded the V.C. for bravery during the famous Charge at Balaclava.

No serious friction between the members of these groups, if indeed there was any, appears to have been brought to official notice. There was however another officer, a cornet named Revell, who fitted into neither of these categories and was, according to Sir Hugh Rose, 'a half-caste'. This young man had become the target for chaffing from his fellow junior officers, which their seniors, far from restraining, seem at times to have encouraged.

In addition a good deal of horse-play and banter had been going on for some time, though probably no more than occurred amongst any other body of young men similarly placed. Towards the end of 1859, two particular incidents were brought to the notice of Sir Hugh Rose by some (apparently) anonymous informer.

The first took place at dinner in the Officers' Mess. In the course of an argument, Captain Thesiger became so annoyed that he picked up some fruit and threw it at Cornet Revell. Whether his aim was good is not recorded, but the junior replied spiritedly with 'obscene and offensive language'.

Thesiger was also involved in the second incident. On this occasion a 'difference' arose between him and a lieutenant named Hardy during some form of gambling. Once again gross and obscene language was used and tempers flared to the pitch where a fight developed. It ended abruptly with the Captain knocking out the lieutenant. To make matters worse, a Captain Rawlinson who witnessed all that happened, did nothing to stop it and failed to report it.

When these incidents were retailed to Sir Hugh Rose, he ordered a Court of Inquiry to look into them. The Court having found all the officers involved to be culpable, Sir Hugh then had Rawlinson, Thesiger, Hardy and Revell placed under arrest and referred the matter to Sir Henry Somerset, the C.-in-C., probably expecting to have them court-martialled.

Sir Henry, however, was not disposed to go to extremes, even in support of his illustrious subordinate. Instead he sent out a Memorandum based on the findings of the Court and the advice received from Sir Hugh himself, in which he severely reprimanded the four officers and indicted Colonel Shute and all the other officers of the Regiment.

The gist of what he wrote is as follows:

'. . . a system of chaffing or personalities on the part of some officers of the Inniskilling Dragoons has led to the serious charges preferred against Captain Thesiger, Lieutenant Hardy and Cornet Revell.'

'. . . Cornet Revell's character has been rather harshly represented by Colonel Shute when he states that Cornet Revell has been habitually coarse and vulgar in his expressions, as the evidence before the Court proves the contrary.'

'. . . Cornet Revell appears to have been made a butt of by his brother officers and did not take the proper means of redress.'

'. . . there has been an absence of propriety and a want of supervision at the Mess of the 6th Dragoons for some time.'

But what upset Sir Henry most about the rumpus was the 'gross and obscene language' used by the delinquents. This, he said, was so disgusting and repugnant to the feelings of a gentleman that if such a word was ever again applied to any officer, that officer must immediately place the offender in arrest when Colonel Shute must see to it that charges of disgraceful and ungentlemanly conduct were preferred.

However, he did not think it would be to the good of Her Majesty's Service if Revell was dismissed or Thesiger and Hardy court-martialled.

Sir Hugh Rose was instructed to read out the Memorandum to the assembled officers of the Regiment and to inform them that they would be punished by having their Mess closed at 11 pm until further notice.

This leniency, which was how Sir Hugh regarded it, failed to have the salutary effects hoped for. A few days later a letter written by Colonel FitzWygram, in his capacity of temporary C.O. in the absence of Colonel Shute, appeared on the desk of

Sir Hugh Rose. Couched in the most polite and correct terms, it was none the less a protest against the comments of the C.-in-C. And while there was no suggestion that the writer had consulted his officers as to its contents, there can be no doubt that it accurately represented their feelings.

Sir Hugh's indignation at this affront to his authority may be imagined. But in dealing with so senior an officer as Fitz-Wygram he decided that a severe reprimand and the withdrawal of the letter would have to suffice. Perhaps, too, he may have taken some consolation from the fact that he would be inspecting the Inniskillings a few days later.

It can hardly be a coincidence that his report on the Inspection should contrast sharply with all the others for many years before and after.

The officers lacked a proper knowledge of their duties— several did not ride well, and the attempt of one to pass a test was 'so bad as to be discreditable'.

The great deterioration in the efficiency of the Regiment, Sir Hugh noted, coincided with the disgraceful conduct of some of the officers. He then detailed the instances already described, and went on:

'I am bound to state that in my opinion, these transactions would not have occurred if Colonel Shute had acted with impartiality and judgement. He entertained a mistaken idea that the credit of his regiment would be increased by its being composed as much as possible of officers of birth, or of a good social position; and he certainly gave way to a tendency to show more countenance to officers of his regiment of this rather than of a humbler class. By so acting he forewent the advantages which result from a commanding officer making military merit the sole claim to his favour; and he lost the personal influence, which as much, and often more than official authority, enables a commander to prevent misconduct amongst his subordinates, and to settle, without heart-burnings or unfavourable notoriety, differences which will occur in all societies.

'I have advised Colonel Shute in this sense; and it is due to him to say that he listened to my instructions in a good spirit, and stated that he would be guided by them.'

Concerning these admirable sentiments it will be observed that Sir Hugh confirmed the suspicion that Shute was class-conscious. At the same time all the evidence suggests that he was popular with his officers (including Weir) and men—even idolized by some. Not that this is surprising since British soldiers traditionally prefer to be ruled by 'a proper gentleman', much to the annoyance of egalitarians.

That Sir Hugh's well-intentioned efforts, far from improving matters, were uniting the Regiment against what they considered as unwarranted interference with their domestic affairs, had been made clear by FitzWygram's insubordinate letter. Further evidence of this appears in the Inspection Report. For though Colonel Shute may have listened politely to his superior's advice, he refused to be drawn into condemning his officers. Asked the routine question whether they were giving him full support, he replied in the affirmative, and this, with obvious reluctance, Sir Hugh was obliged to record.

It is also worth noting that although there had been apparently a disastrous fall in the morale and efficiency of the officers of the Regiment there seems to have been no such decline amongst the N.C.O.s and men, whose behaviour had been the one feature to meet with Sir Hugh's disapproval in an earlier report.

Now, it seems, an improvement had actually taken place. In public the soldiers had 'a clean and military appearance', and 'the behaviour of some old and hardened offenders had changed for the better, and no instances of misconduct of any soldier of the regiment towards the inhabitants, or other soldiers has come before us'.

The inconsistency of a high level of soldierly qualities in the rank and file when their officers were said to be in a state of social anarchy and military inefficiency was to be a recurring theme in the history of the 6th Dragoons for some time to come.

Not surprisingly perhaps, Colonels Shute and FitzWygram now made up their minds to leave the Regiment and return home to England. The decision could not have been easy. Shute had been with the Inniskillings for 20 years, FitzWygram for 17. Of the reasons advanced for their departure one was that the Regiment was soon to move up country to Ahmednuggur, while many years later it was said of FitzWygram that he exchanged to the 15th Hussars because there seemed no prospect of active service in India, though this seems implausible as the scene in England was equally peaceful.

The two seniors were not the only ones to quit. The disturbed state of the Regiment, the move up-country, the impending departure of their C.O., all combined to persuade a large number of other officers to make plans for returning home.

Their arrangements took many months to complete, however, so it was with virtually the same personnel that the Regiment was again inspected in October, 1860. Sir Hugh Rose had left in May for Calcutta to take over the command of the Army in India and the inspection was carried out by a Major-General Hall. The contrast between his report and that of only seven months before must have been some consolation to Shute in particular, and is quite remarkable.

It begins: 'The state of the regiment is most satisfactory in every respect', and goes on to lavish praise in detail. 'The commanding officer, Colonel Shute, is a most zealous and talented officer, and peculiarly well qualified for training cavalry. His attention to his duties since he has served under my command has been unremitting, ably and zealously seconded by Lieutenant-Colonel FitzWygram. Since the last inspection . . . nothing has occurred to form an exception to the very favourable report which I have the satisfaction in making, either as regards officers or men.

'Captain Thesiger, whose conduct was remarked on by my predecessor, Sir Hugh Rose, at his last half-yearly inspection, was detached in May to China with horses, and has not yet rejoined. Cornet Revell is also absent on a similar duty . . .'

Perhaps Thesiger and Revell managed to improve their personal relationship en route. They certainly widened their military experience as both took part in the China War and were present at the taking of Pekin. The move to Ahmednuggur was completed in November. Soon after, the officers, including FitzWygram and Hunt, began to leave for home. Colonel Shute was one of the last to go, in March, 1861.

After his departure, six weeks were to elapse before the new C.O. arrived from England and in this period the 6th Inniskilling Dragoons were commanded by their senior Troop Captain, Archibald Weir. Whatever qualities the Captain may have brought to his unexpectedly high office, he at least showed a determination to tread the path of his immediate predecessors.

Eleven days after Colonel Shute left, the Brigade Commander, Brigadier Hobson ordered a full dress parade of the Regiment. He found the Sergeants' Mess, the Canteen and the Library dirty and wrote to Weir expressing his dissatisfaction with what he had seen.

On receipt of this communication, Weir proceeded to consult with his officers and wrote back disagreeing with the Brigadier's derogatory remarks. He then went even further than Fitz-Wygram had done and refused to apologize or to withdraw his letter when ordered to do so.

Yet strangely enough his punishment was no more than a severe reprimand and a solemn warning against repeating the offence.

Once again the Inniskilling Dragoons had, in the words of Sir Hugh Rose, shown 'impatience of reproof, unaccompanied by amendment'.

3

A Social and Moral Question

In exchanging to the 6th Inniskilling Dragoons, Colonel Crawley could have had little idea of the problems he was to inherit occasioned by the departure of the field officers, the large turnover of other officers and above all the strained relations that had developed between those who remained and superior authority. Nor was his task to be made easier by the reputation for strictness that, by some telepathic process, had preceded him.

The details of his arrival on the morning of 25 April, 1861 were later much disputed, but his reception was probably not over-cordial. In the evening he met all his officers and informed them he intended to be strict on matters of discipline and military efficiency, but that off duty he hoped to meet them on the most friendly terms, which, as an avowal of intentions, was unexceptionable.

Evidence of how he proceeded to put the first part of his policy into effect is to be found in the numerous General Orders posted in rapid succession and in their wording.

In May, for instance, came strict instructions that the bedding of horses was not to be taken away immediately, but raked back, so that manure could be shaken out and the good bedding material left. 'The C.O.', it warned, 'will soon have an opportunity of judging each troop by the increased quantity of bedding, how it has been obeyed.'

Later, on the same subject, sentries were ordered to be on

duty at all times 'to prevent good bedding being taken away to feed the contractors' buffaloes'.

A June Order announced that a Corporal had been reduced to the ranks for 'screening' drunkenness in a man on duty. All N.C.O.s were warned to 'beware of how they tamper with discipline by such unworthy and disreputable conduct, which will be sure, if discovered, as it infallibly must be, to lead to their degradation and disgrace, and to the door of promotion being effectively shut in their faces forever.'

As these samples indicate, Crawley had a penchant for the highflown phrase and the veiled threat.

Whether he would eventually have overcome his difficulties and moulded the Regiment to his own exacting standards of efficiency and discipline cannot be known. For the struggle for ascendancy in the regimental sphere was soon ousted to second place by a more pressing problem in the social life of the little community. And here again Crawley may be accounted unfortunate, for the seeds of the conflict had lain dormant within the Regiment for more than a year and only broke the surface soon after his arrival. It was first brought to his notice in an atmosphere of deceptive innocence on an evening, early in July, 1861, when he received a visit from two of his senior departmental officers, Captain Smales the Paymaster, and the Chief Surgeon of the Regiment, Dr Turnbull.

Smales, the elder, was a tough campaigner of 50, with a long and varied military career stretching over thirty years. The son of a sea-captain, he had evidently inherited his father's roving disposition. His first connection with the Army was as Acting Ordnance Clerk in Sierra Leone in 1830, a spot where the mortality rate for Europeans hovered round the 75% mark. Soon after, he became a regimental officer and saw service in one of the Kaffir Wars.

In Freetown in 1842, he married a young lady euphoniously named Clara Sarah McDonald. About the same time he changed to his present occupation and was paymaster in several cavalry regiments, spending in all eighteen years on the West Coast of Africa before returning to England.

During the Crimean War he was Chief Paymaster of the Sanitary Commission which was sent out to Scutari to investigate the conditions under which the British Army was living and which, according to Florence Nightingale, saved the Army in the Crimea.

He joined the Inniskillings soon after they arrived in India and appears to have gone about his business quietly and efficiently without being in any way involved in the events which had brought its officers into disputes with Authority. He now had a family of five young sons.

His companion, Dr Turnbull, was a more stolid individual in his middle thirties.

They began by assuring the Colonel that they had called purely out of a sense of duty and as an act of courtesy. A disquieting rumour was circulating in the station. It was alleged that the wife of a troop captain named Renshaw was a woman of the most immoral character whose first husband had divorced her about three years before in circumstances of the utmost depravity. It was also said that Captain Renshaw had been involved in the scandal as her lover.

They had consulted the other married officers and decided they could no longer meet the 'lady' on the same intimate footing as before and would in future avoid all social contact with her whenever possible. Then, according to the account they gave later, Crawley thanked them for telling him, asked to be kept informed of anything more they might hear on the matter and took leave of them on the best of terms.

How long the rumours about the Renshaws had been circulating is uncertain but at this stage no one seems to have known precisely what had happened. Their curiosity was soon to be satisfied, for Smales had sent to England for copies of *The Times* of 13 and 14 May, 1858, which gave a full account of the divorce proceedings.[5]

Why he should have done this suggests a more personal motive than his professed moral indignation. A reason advanced later was that the instigator of the social ban on Mrs Renshaw was Mrs Smales who had taken a dislike to the younger woman. And as the divorcee was 'a lady of great personal attractions', this sounds plausible.

In the days which followed, Crawley was clearly perplexed as to what course to take. In ordinary circumstances he would probably not have dissented from the conventional view on the place of a divorced woman in society. But the changes he had introduced into the running of the Regiment, and especially his manner of doing so, had alienated all his senior officers except Renshaw. And while he had been on friendly terms with Smales and Turnbull up to this point, it was hardly worth sacrificing his ally amongst the regimental officers for these two.

Moreover, the Renshaws had been recommended to him by Colonel Shute as a charming couple and they had since become very friendly with his wife and mother-in-law.

The more he thought of it the more it seemed that this 'social and moral question' being presented to him at this particular time was no mere accident. The fact that the married officers had consulted together began to appear as an attempt to dictate to him a particular line of conduct and part of a plot to undermine his authority.

For some weeks he took no definite action but his 'reserved manner' left Smales and Turnbull in no doubt as to his displeasure. Then, on 22 July he disappeared on a week's leave to Poona where he called on Sir William Mansfield,[6] Commander-in-Chief of the Bombay Army and gave him an account of his difficulties, naming Smales and Turnbull as active trouble-makers.

He also requested that Colonel Smith, a cavalry officer, should be sent to Ahmednuggur to put the Inniskillings through a thorough inspection and at the same time do what he could to smooth out the 'misapprehensions' that had arisen between himself and his officers. To this the General willingly agreed.

But then, coinciding with Crawley's return to the Regiment, an incident, trifling in itself, served further to widen the rift.

The single officers were giving a dinner party to the community at the station and must have decided to extract some amusement from the 'social and moral question' that had aroused such strong emotions in their elders. When Turnbull arrived at the Mess he was accosted by a very junior cornet who said with a smirk, 'Oh Turnbull, you'll take Mrs Renshaw to the dinner table.'

The surgeon replied he would take any lady he liked.

The cornet retired, but returned in a few minutes with, 'Oh

Turnbull, you'll be sitting next to Mrs Renshaw at the Mess-table.' To this Turnbull replied that he would sit where he pleased—which he proceeded to do.

After the party, several officers assembled in the ante-room where an argument developed between Turnbull and Thesiger, the President of the Mess, as to whether the former had been right in refusing to take in to dinner the lady who was told off for him. Remarks were passed that continued to rankle after the contestants had parted for the night.

The following morning details of what had transpired were reported to Crawley by Thesiger and both he and Turnbull were summoned to the Orderly-Room. Here Turnbull contended with a good deal of heat that Thesiger had no right to speak to him as he had done before junior officers and accused him of abusing his position as Mess President.

It was clearly a situation calling for conciliation. But Crawley had now become so committed on the thorny subject of the Renshaws that he could not resist the opportunity of putting one of his prime opponents in the wrong. He joined the argument on the side of Thesiger and ended up by reprimanding Turnbull for his conduct.

As might be expected, his partisanship served only to make Turnbull and those who shared his views all the more determined to oppose him in the future. The 'social and moral question' had developed into a matter of principle on which neither side was prepared to budge.

A few days later, the Surgeon counter-attacked with a letter in which he reiterated the unanimous opposition of the married officers to meeting Mrs Renshaw socially and pointed out the desirability of Crawley's taking the lead in excluding her from the social life of the Regiment.

Crawley wrote in reply that it was his duty to protect the interests of all under his command and to promote harmony in the interests of the Service. He then went on to make the issue a personal one, defending his decision and allowing, as it were, his inner uncertainty to leak out.

'It is wholly unnecessary for anyone to dictate to me the line of conduct I shall pursue in the matter under consideration or in any other. I shall be guided by the principles which have never failed me during my years of service, and as these principles are justice and moderation, I have little fear they will carry me through as they have hitherto done.'

A few days later he summoned all his senior officers to his house. Smales and Turnbull, he said, had been spreading malicious rumours about a brother officer with the object of causing disharmony in the Regiment. But anyone engaging in that sort of activity should first be sure that their own past would stand up to scrutiny.

Then, pointing dramatically to a tin box on the table, he announced that it had been left him by Colonel Shute for just such an occasion as this. 'Colonel Shute's Legacy', he said, contained a record of the social and moral lapses of all Inniskilling's officers that had come to the late C.O.'s notice, and if the contents were to be disclosed, few of those present would be able to point the accusing finger at anyone.

He rounded off the meeting by directing that there must be no further reference to the subject to him and no further discussion of it by any officer under his command.

The introduction of the 'Confidential Box', obviously intended to intimidate the opposition, had the immediate effect of provoking a counter-move. Turnbull and Smales wrote asking to be supplied with copies of any adverse remarks on their past conduct it might contain.

Meanwhile Colonel Smith had arrived, ostensibly to inspect the Regiment—ostensibly, for it would probably be to underestimate the intelligence of Crawley's opponents to suppose them incapable of making a shrewd guess at the real purpose of his unscheduled visit following so closely on Crawley's absence in Poona. Nor was it likely to heighten their respect for their Commanding Officer.

Yet it does appear that the Colonel had some temporary success in relaxing the tension. Smales and Turnbull obtained an admission that their names did not appear in the 'Confidential Box'. They expressed themselves satisfied with the explanation and said they would make no further representations on the subject. On the 'Social and Moral Question', however, they insisted they must remain free agents and be allowed to act as they thought fit.

One other incident is worth noting. As the officers were leaving, Colonel Smith called Smales back to warn him about his highly objectionable manner, which, should the occasion arise, would be remembered and reported on.

There followed a truce, at least in overt acts of hostility. But the calm was illusory. Far from relations improving, the rancour already aroused was becoming more firmly established. And in the atmosphere that now existed Crawley on the one side, and his two chief opponents on the other, began to sense that all-out war was becoming inevitable.

Social contact between Crawley and the married officers (except Renshaw) had virtually ceased. Yet all were forced to meet and carry on the day to day business of soldiering. It required only some provocation, real or imaginary, to provoke retaliation and for the life of the Regiment to be thrown completely into confusion.

The first of what were to be a series of incidents accompanied by mounting tension occurred in October, 1861.

A number of men were in camp for musketry practice. The weather was unusually wet, and Turnbull, after visiting them, recommended they should be sent back to barracks.

On his return, Crawley sent for him and demanded his reasons for making this recommendation. Turnbull explained that the wet conditions laid the men open to fevers and bowel complaints.

Crawley disagreed. It was time to send the men back when they became ill.

'No,' said Turnbull, 'prophylactic measures are generally taken into consideration nowadays.'

Crawley said he would bear the suggestion in mind but asked by what authority Turnbull had interfered in the matter?

Turnbull replied that he was obliged to do so by Army Medical Regulations which he would send along for Crawley to consult if he wished. He added that he was also obliged to report any recommendations made to his C.O. to the Deputy Inspector of Hospitals.

In his report he supported his case by quoting Brigadier Hobson with whom he had discussed the matter on the spot, for that officer had concurred with him that it would be better if the men were not unnecessarily exposed to the rainy weather.

But in this Turnbull was ill-advised. Higher authority would be unlikely to view with favour the use of 'off the record' remarks by a brigadier to strengthen the case of a subordinate against his C.O.

Inevitably Sir William Mansfield, before whom the correspondence eventually arrived, delivered the Surgeon a stern reprimand for making unauthorized and unwarranted use of Brigadier Hobson's remarks, adding that his conduct was more calculated to cause disunity than promote the health of the men.

It had now been made abundantly clear to both Smales and Turnbull where higher authority stood in the quarrels between them and their C.O. and that should they persist they would land themselves in very serious trouble.

4

'A very fire-brand amongst inflammable material'

While Turnbull appears to have been most active in his disapproval of the Renshaws and while many regimental officers had been at cross-purposes with their C.O. on other matters, by the end of October, Crawley had come to the conclusion that his most dangerous opponent was Paymaster Smales.

The explanation for this is partly revealed in the disapproval of Smales' manner by Colonel Smith following the 'Confidential Box' incident. Uncertain of how best to wield his authority, Crawley was undoubtedly, and with reason, apprehensive of this contemptuous and inflexible subordinate. So much so, that his heightened imagination had now identified the Paymaster as the mainspring of all the opposition which sprang up against him in so many quarters and had convinced him that his problems would not be solved until he had rid himself of this trouble-maker. And if he could be removed from the Regiment in such a way as to serve as a warning to the rest, so much the better.

But to accomplish this objective a pretext must be found and therein lay the difficulty. Smales was far too experienced and shrewd to make mistakes in the course of his duties likely to provide the chance of bringing against him charges of incompetence or infringing regulations. Besides, to wait for such an unlikely eventuality was not expedient.

The alternative was to provoke him into some indiscretion on which a charge could be built. But here too Crawley was at a disadvantage. He needed reliable accomplices and these were in short supply. The attempt must however be made and at the beginning of November he decided to make it. The occasion was the half-yearly inspection of the Regiment.

On the evening of the 6th, as Smales was dressing for the dinner at which Major-General Hale, the Inspecting Officer, would be the guest of honour, he was surprised by the sudden and unexpected appearance at his house of Captain Renshaw in full dress uniform and obviously on his way to the Mess.

The call could not conceivably be a friendly one. But the Paymaster's emotions may be imagined when Renshaw said he had come to draw the money due to him for his troop account. To reinforce his demand he produced a curt letter from Crawley directing that the cash be issued 'immediately'.

Smales, who must at once have recognized an *agent provocateur*, refused, pointing out that the money should have been collected, if in fact it was due, before 4.30 pm. and that it was not possible or indeed within regulations to issue it at that late hour. As soon as Renshaw had gone he despatched a short reply to Crawley declining to make the payment.

Ten minutes later he received a second caller, Captain Weir, also dressed for dinner, and with a communication even less palatable than the first—he was to consider himself under arrest for refusing to obey an order from his Commanding Officer.

Now whether this course of events was actually planned by Crawley or not, the result was certainly what he wanted. Not only was Smales shown up in the most unfavourable light before General Hale, he was also denied all communication with him while Crawley was able to voice any complaints he wished without danger of contradiction.

The General's comments in his Report show how effective the plan had been: 'Paymaster Smales (whose conduct is the subject of a Court of Inquiry), is reported by Lieutenant-Colonel Crawley to be insolent and insubordinate, and by his litigious disposition to have caused disunion and bad feelings throughout the regiment, and to be "a very firebrand amongst inflammable materials". Whatever may be the result of the inquiry on Paymaster Smales' case, it appears certain that so long as he remains in the regiment, there is no probability of a good understanding

3

being restored to the regiment and I therefore strongly recommend the removal of Paymaster Smales from it.'

Two other officers, Captain Swindley and Surgeon Turnbull were accused with Smales as the chief instigators of the discord. These two were seen by the General and warned of the dangers they ran by going against their C.O.

At the same time Crawley himself did not escape criticism. The Report described him as 'a zealous and intelligent officer, but wants tact and temper, especially in his relations with his officers'.

As might perhaps be expected, the solemn warnings of the General and the sudden arrest of Smales did nothing to limit the hostilities. On the contrary, Crawley's abuse of his authority raised tempers to such a pitch that behaviour on both sides now achieved further heights of absurdity.

A whole day passed before Crawley realized that by placing Smales in close arrest, he had left the Regiment without a paymaster and effectively immobilized the pay office. For the existing regulations required a paymaster to provide securities before being appointed and to safeguard these he had the right to nominate a substitute to carry out his functions at any time.

When Crawley did get round to calling on Smales to name another officer to stand in for him, the method he used was both unusual and vindictive. He directed him to do so in Morning Orders at the same time informing the Regiment that he had been arrested for refusing to obey his Commanding Officer's orders.

The findings of the Court of Inquiry set up to decide on the circumstances of the arrest duly arrived at H.Q. in Bombay before Sir William Mansfield. The General's reaction was immediate. He despatched a message by telegraph ordering the release of Smales and directing that no further action must be taken pending his decision.

This arrived by post a fortnight later, and showed the considerable strain imposed on the General in attempting to reconcile his support for the C.O. with that of elementary justice for the Paymaster.

Great mistakes had been made. Colonel Crawley ought not to have placed the Paymaster in arrest. Before taking so serious a step he should have reflected on the lateness of the hour at which the money was demanded since strict adherence to opening and closing times was essential for the security of the Pay Office. He should also have taken into consideration the important position held by Captain Smales in the Regiment and the fact that he had always carried out his duties to the complete satisfaction of his commanding officers including Colonel Crawley himself.

Before ordering the arrest he should have pondered on how he proposed to keep the Pay Office functioning. Only in cases of defalcation ought Paymasters to be placed under close arrest. For breaches of discipline the obvious course was to place him in open arrest so that he could carry on his normal duties in much the same way as a medical officer would in the same circumstances.

Captain Renshaw obviously went to his C.O. with a grievance which did not exist. He must have known that it was against Regimental Orders to ask for money from the Paymaster at 7.15 pm. In any case it was now proved he had already drawn ample money for his needs.

Lieutenant Davies (a subaltern in Renshaw's troop), was labouring under a gross misapprehension in stating that Captain Smales broke arrest by deliberately riding round the camp and ostentatiously appearing in front of the C.O.'s house. This assertion was denied by the prisoner and disproved by other witnesses. Captain Smales was hardly likely to have been guilty of such bravado at a time when he was in such serious trouble.

As to Paymaster Smales' contention that Colonel Crawley had no power to place him in arrest because of the conditions under which paymasters take service, this could not be accepted. He was a military officer, subject to the Articles of War and military discipline like any other.

Sir William trusted this would be the last he would hear of dissentions within the 6th Dragoons and that 'concord and forbearance' would be the rule from now on. He also desired there should be no further correspondence on the subject.

Before this arrived, however, Smales, upon his release and smarting at the indignities thrust upon him, had plunged into a frenzy of activity. He attempted to press charges against the officers who had given evidence against him and threatened

Crawley with a civil action for placing his sureties at risk. (In the light of which even the 'bravado' disbelieved by Sir William Mansfield seems probable.)

Then, when the General's ruling on his arrest came, more rationally, and no doubt spurred on by the fact that it was largely in his favour, he set about writing a long Memorandum giving his reasons for the strained relations between Crawley and himself, and contending that his refusal to pay out money to Renshaw had been the pretext for his arrest while the reason was the disagreement over the social and moral question.

His treatment by Crawley had been most unjust and uncalled for. He had been arrested on a matter 'looking very like disobedience of orders', had been the subject of an Official Confidential Report to Major-General Hale, and his name had been paraded in Regimental Orders for no other purpose than to disgrace him in the eyes of the Regiment.

For these reasons he humbly and respectfully requested Sir William Mansfield to grant an enquiry into the propriety and legality of Crawley's conduct towards him.

While the Paymaster was feverishly preparing his Memorandum and pressing for that and other correspondence to be forwarded to higher authority, the Regiment too was in a state of unusual activity, being engaged in the long and arduous process of moving from Ahmednuggur to Mhow, a distance of some 500 miles. It was therefore in Crawley's possession for about three weeks before he forwarded it from his new station to Bombay.

His covering letter expressed a good deal of injured innocence. He deeply regretted having to trouble His Excellency with the 'voluminous correspondence' forced upon him 'by the importunity and repeated demands of Paymaster Smales', who, 'with all the courteous defiance and respectful innocence for which his correspondence is remarkable, insinuates the most unworthy motives against me, and a dereliction of duty as flagitious as could well be imagined in a Commanding Officer, in being biased against him and others who have given me offence. . . .'

In little more than a week Smales was made painfully aware that far from enlisting support for his case, his Memorandum had made his situation even more precarious. For Sir William

Mansfield had by now received Major-General Hale's recommendation that the first step in restoring the authority of Crawley was to remove Smales from the Regiment. His reply to the appeal included a stern reprimand to be read out to the Paymaster by the Officer Commanding Troops at Mhow in the presence of Crawley.

It pointed out that not only had Smales attempted to prefer charges against officers not under his authority but had done so when Sir William was considering the proceedings of the Court of Inquiry on the differences between them. This, to use the mildest term, was a breach of military decorum. He had also used the argument of his sureties to deny the legitimate authority of his Commanding Officer.

For Smales the implications of this repulse were all too clear. There was now no one in India to whom he could turn for redress and he was virtually at Crawley's mercy. He could accept defeat, but to a man of his background and temperament, capitulation was unthinkable. The alternative was to attack, for to let matters drift was to await destruction. But how to attack one so impregnably placed as Crawley appeared to be?

There was only one point at which he might have the advantage—the volume of support he could expect from within the Regiment if a direct conflict between him and his C.O. were to arise, and as events were to show, it was this that prompted his next move.

On 8 January he wrote directly to the War Office notifying the Secretary of State for War that Crawley had on several occasions failed to attend Muster Parades. In doing so he was, of course, breaking the regulation that all communications to higher authority must be sent through the commanding officer, and what he hoped to achieve is hard to say. Perhaps he was trying to signal the unhappy state of the Regiment to the War Office to create the opportunity of putting his case, even if this brought a reprimand.*

* The significance of the reference to Muster Parades was that all officers on the strength of the Regiment, not on leave, sick or otherwise officially engaged, were required to be present on the first of the month when they took place. Absence meant, technically at least, that the officer concerned could not draw his pay for that month. Smales as Paymaster was officially involved so that his request for advice as to whether Crawley ought to be present gave his letter a slender claim to validity.

But as the weeks passed and Crawley's intention to get rid of him became daily more unmistakable, the hope that help might come in reply to his letter faded and he determined on a more desperate attempt to bring matters to a head.

Just how desperate he could hardly have known. For it was one that was profoundly to affect not only his own life but those of his comrades and of many others who at the time knew nothing of his existence. On 26 February he addressed a letter to his Commanding Officer.

His position, he wrote, had now become so painful that he felt bound, respectfully, to seek the protection of higher authority.

He went on to protest that he had done all in his power to carry out Sir William Mansfield's injunction that 'concord and forbearance in the discharge of duty' were necessary for the good of the Service but had been met only with hostility in return. He then detailed instances of how he had been victimized.

Because he had relaxed certain rules to oblige officers, including Crawley himself, he had been reprimanded by the Major-General at Mhow. This, he maintained, must have been due to verbal representations by Crawley.

Crawley had made written statements of conversations with him in the Orderly Room, and other officers present had also been told to write statements for use 'as they might be required'.

He had applied for and been granted a month's leave and had named an officer to perform his duties in his absence, when the Major-General had suddenly made his departure conditional on a certain account being passed by the Audit Department. At the rate business was usually transacted his leave would have expired before the first reference was made to the account and in any case, Captain (now Major) Swindley was perfectly competent to deal with the matter. If all Paymasters were treated in this way, they would never be granted leave in India.

Now had Smales concluded his letter at this point he would probably have placed himself in little danger. But equally, as he well knew, his appeal for justice would have achieved nothing. He therefore went a good deal further, moving from defence to attack, from the role of accused to accuser.

Crawley, he said, had severely scrutinized the way in which he carried out his duties and for the most trivial and unintentional errors had accused him of infringing the Articles of War, but had not applied the same standards to his own conduct. For although it was clearly laid down that the commanding officer of a regiment must be present at the monthly muster parade he had been systematically absent.

He had been absent from the first parade after he assumed command of the Regiment and from the first parade on arrival at Mhow. Yet on these and other occasions he had signed himself present on the Adjutant's Roll. 'What,' asked Smales, 'would my position be had I acted in such contravention of the Articles of War.'

In addition, some of Crawley's remarks to his officers had been anything but conciliatory. He had expressed *regret that the days of duelling were not in existence*, sentiments hardly likely to establish 'concord and forbearance', while his 'harsh and unusual tone and style of address' left little hope that where offence had been committed apology or explanation would prove acceptable.

Smales ended as he began by regretting the need for these comments, for in all his long service he had never had 'a shadow of misunderstanding with a commanding officer'. But he had been treated with such hostility that he had no alternative but to seek protection by requesting that his communication be forwarded to higher authority.

For two days after writing his letter he pondered and consulted his friends on the advisability of despatching it. Then on 28 February at 2.30 pm he handed it in at the Orderly Room.

5

Hobson's Choice

Crawley later complained that for the first eight months he commanded the Inniskillings there were no field officers serving with the Regiment with whom he could discuss the many difficulties that beset him.

Lieutenant-Colonel Hunt, nominally his second-in-command, having failed to obtain the command of the Regiment, had sailed for England before Crawley arrived and remained there on extended leave up to the end of 1861 when he retired from the Army altogether. He sold out to a Major Prior of the 12th Lancers who arrived at Mhow just before the feud between Crawley and Smales came to a head with the Paymaster's insubordinate letter.*

On his journey, Prior had passed through Poona where Sir William Mansfield earnestly requested him to do all in his power to smooth out the troubles between his new C.O. and the officers. This for any new-comer would be a tall order. But for Prior matters were complicated by the fact that he already knew two of the officers most strongly opposed to Crawley. He had served with Swindley for over six years in the 12th Lancers and they were on active service together in Africa, the Crimea and the Mutiny; he also knew Turnbull who had been attached to the 12th in the Crimea.

In this unenviable situation, Prior would obviously need

---

* The Major of the Inniskillings was similarly absent and sold out at the same time to Captain Swindley.

unusual discretion and detachment, qualities in very short
supply among Inniskilling Officers at the moment.

His first entry into the struggle between the two chief antagon-
ists was an attempt to persuade Smales to withdraw his letter;
though what exactly took place can only be surmised from the
contradictory statements by Smales and Crawley. Prior wisely
avoiding involvement by publicly saying nothing about it.
Smales contended that Prior had been sent by Crawley, and
though Crawley indignantly denied this it would certainly have
suited him if Smales could have been persuaded to take back his
letter unconditionally. He would then have been relieved of the
need to take action against his opponent and so have avoided the
uncertainty and unpleasant publicity that must follow.

For once the letter reached higher authority, the next step
would normally be the setting up of a Court of Inquiry at which
the Paymaster would be able to bring forward all the evidence
he could to support his accusations. Indeed, after his experience
of the previous November, this was precisely what Smales was
hoping for; it was also what Crawley wanted to avoid. At all
events Prior's attempt at appeasement failed.

For Crawley this was a set-back. But by this stage his fortunes
had, in one respect, taken a turn for the better. Since the move to
Mhow he had acquired a staunch and valuable ally in the Garrison Commander, Major-General Farrell, who according to one
description was 'an aged and infirm Indian Army Officer, mostly
confined to his couch'. Certainly the General's physical and
probably his mental condition—he was 62 and had joined the
Bombay Army way back in 1818—placed him very much under
Crawley's influence. He was unable to get around to see exactly
what was going on but Crawley would call on him to give his
own highly personal view of events. Some indication of the
assistance Farrell was prepared to give is already apparent from
the reprimands he had handed out to Smales and the sudden can-
cellation of his leave.

The problems posed by Smales' letter were earnestly discussed by these two before any decision regarding it was taken.
Then after a delay of over a week it was finally decided to send
it on to Sir William Mansfield with a covering letter by Crawley
vigorously denying and refuting all the accusations it contained.

'These are accusations,' he wrote, 'which might well make a man tremble, but which I am happy to acquaint the Major-General and his Excellency the Commander-in-Chief contain not one grain of truth. . . . I charge Paymaster Smales with making a series of false accusations against me, his Commanding Officer, as detailed in his letter herewith forwarded, and I respectfully call upon his Excellency to give me the opportunity of proving that they are false, and of thus vindicating my character to the world.'

To all this Farrell lent his full support. The complaints by Smales, that he was being persecuted by his Commanding Officer were he said, completely without foundation, while the accusations that Colonel Crawley was infringing the Articles of War were of so serious a nature that he was framing charges against the Paymaster.

He concluded, 'It does not appear to me that in the present instance a Court of Inquiry would be necessary, nor do I conceive it would be advisable under the circumstances to place Paymaster Smales under arrest pending instructions from Army Headquarters.'

Obviously Crawley had by now learnt caution in dealing with his wily opponent. This time there would be no hasty arrest followed by a Court of Inquiry.

A week later the reply came from Bombay. Sir William Mansfield agreed that Captain Smales should be brought to a Court-Martial without any preliminary inquiry into the merits of the case. He went further, and in what can only be interpreted as an attempt to make the Paymaster's destruction doubly sure, appointed Crawley to be the Prosecutor at the trial.

This extraordinary step of course placed Crawley at an enormous advantage. He would not only be able to present and argue his own case before the Court but could offer himself as a witness on oath for the prosecution.

Probably Sir William's chief desire was to end once and for all the dissentions that were disrupting the Regiment by firmly establishing the authority of its C.O., and the fact that he was tampering with the machinery of justice to achieve this end may or may not have troubled him. But, in doing so, he had set foot on a slippery slope, for he had staked his reputation on the word of Colonel Crawley.

6

The Case for the Prosecution

The Court-Martial on Paymaster Smales began on 1 April. It had as President, Lieutenant-Colonel Payn of the 72nd Highlanders (Seaforths), with four lieutenant-colonels, one major and nine captains as members.

The Charge read: 'For conduct highly insubordinate and most disgraceful and unbecoming the character of an officer and a gentleman and to the prejudice of good order and military discipline in the following instances: In having in an official letter dated Mhow, the 26th day of February 1862 to the address of Lieutenant-Colonel T. R. Crawley, his Commanding Officer, made the following false and malicious statements with reference to the said Lieutenant-Colonel Crawley, viz. . . .' Then followed the three paragraphs from Smales' letter in which he accused Crawley of failing to attend Muster Parades while signing himself present, of causing discord in the Regiment by his 'harsh and unusual manner' and of expressing regret that duelling was no longer in use for settling differences in the Army.

The proceedings opened with what was to be one of its main features, a protest by the Prisoner. In it he argued at some length on the illegality of the means by which he had been brought to trial. He claimed that he had not been allowed to bring evidence to support his accusations at a Court of Inquiry; that it was illegal to quosh the charges he had made by bringing a counter charge against him; and that this was the second time in a few months he had been arrested on ex parte evidence.

The Court then set the pattern for its future conduct by

43

rejecting the protest outright and in this instance advised the Prisoner in his own interests to withdraw it, as it reflected on the conduct of His Excellency the C.-in-C. by whose authority the Court was convened. Before this obvious coercion Smales agreed to do so for the time being.

Crawley followed with a speech in which he regretted the need for the Court-Martial but said it had been forced upon him as the only means to vindicate *his* character against the malicious accusations of the Prisoner.

He then began the case for the Prosecution by offering himself as a witness on oath and went on to give details of all the Muster Parades from 1 May, 1861, to 1 January, explaining his presence or absence on each occasion. But as the issue soon narrowed down to whether he was present on the two dates mentioned it is enough to give the substance of what he had to say on these.

On 1 May he had been unable to take command of the Parade because his undress uniform had not yet arrived from Bombay. He was however present in plain clothes and watched the troops as they filed past. After this he inspected the officers' horses.

In answer to a question by the Prisoner as to whether he commanded the Parade he said:

'. . . I consider myself the commander of that parade equally in plain clothes as if I had been in uniform. All the details of that parade took place under my eye, the same as if I had been in the full uniform of the regiment; and I venture to think that no officer or soldier of the regiment would have ventured to disobey an order given by me on that parade.'

As regards the Muster Parade of 1 January, 1862, the Regiment was that morning about to move into barracks at Mhow having lived in tents since they completed the march from Ahmednuggur. Crawley said he had spent all the previous day seeing that the barracks were made ready for occupation, and before the Parade he went round with the Quartermaster and the Sergeant-Majors on a final inspection. He then proceeded to the horse lines where the troops filed past him.

In support of these statements he brought forward several witnesses. He said later that he could have produced nearly every soldier on the parade to prove his presence. If so, his selection was surprisingly small and the evidence they were able to give inconclusive.

As might be expected Captain Renshaw and his subaltern, Lieutenant Davies, gave him their support. Davies in an apparent attempt to add verisimilitude to his account described how, on 1 January, he saw Crawley coming towards the parade ground 'near the haystacks'. The Defence however managed to disprove this. For on that day there could have been no haystacks to pass since they were not commenced until weeks later.

Another Captain named Garnett also swore to having seen Crawley watching the troops file past. But his testimony was weakened by the evidence of others who swore that he was not even with his troop when it marched off the parade ground. His troop sergeant in particular recalled having himself marched the men off on account of Garnett's absence.

Further evidence for the Prosecution was given by Veterinary Surgeon Poett (of whom more later), two sergeants conveniently released from arrest for the purpose, and two privates.

Three officers, Captain Weir who commanded the Parade in May, Major Swindley who was Acting Paymaster at the time, and Lieutenant FitzSimon the Adjutant, were of necessity questioned. All swore they had not seen Crawley on that Parade or the one in January. The Quartermaster also stated that he was with Crawley in the Barracks on the morning of 1 January, but saw no troops file past him on that occasion.

An apparently convincing piece of testimony did however come from a subaltern named Bennitt who swore to being on Parade on 1 May and to seeing Crawley there during the time the roll of the troops was being called. He gave the Colonel's position as towards the left front of the column and close to it. He also recalled the troops filing to water and certain remarks passed by Crawley about the appearance of the horses and the length of the hair on their legs.

Before being cross-examined, Bennitt was asked to look at two orders, both dated 30 April, 1861. One showed him as having been on a Watering Parade on that date, the other as having been detailed for a Treasury Board at 6 am on 1 May. He was then asked how he accounted for being present on a Muster

Parade at the time he was ordered for committee duty. He replied he had no recollection of the committee duty but remembered the Muster Parade perfectly.

Questioned by the Court as to why he was so sure, he said he remembered calling Captain Weir's Troop (to which he belonged) to attention when he saw Veterinary Surgeon Poett appear. This was because Poett, like Crawley, had just joined the Regiment and he mistook one for the other.

The following day, however, Bennitt put in an unscheduled appearance to request the Court's permission either to qualify his previous answers with 'to the best of my belief', or else to withdraw his evidence altogether. He had telegraphed Nuggur and been told he had signed certain papers at the Treasury on 1 May, so he must have been there when the Muster Parade was taking place. He must, he said, have confused it with the Watering Parade of the previous day since both were alike except for the roll-call,* and—

'At the distance of a year in time one can only depose to facts upon the best of one's belief and that was all I meant to do, or all I consider I could do, after such a length of time.'

The Court found this explanation quite acceptable, though the Prisoner, understandably took a less tolerant view of Bennitt's lapse of memory. He handed in a written protest which the Court accepted with reluctance:

'We cannot receive your address without expressing our opinion that you have indulged in the most unwarrantable and offensive recriminations on the Prosecutor and the most unjustifiable reflections on a number of the Prosecutor's witnesses; and we would remark more especially on your impeachment of the truthfulness and honour of Lieutenant Bennitt whose explanation has already been pronounced and is still considered by the Court as perfectly satisfactory.'

---

* This was something short of the truth.

7

The arrest of the Sergeant-Majors

By the ninth day of the Trial, when the case for the Prosecution
closed, it must have been obvious to Crawley that in spite of the
outright support from the Court, there was a danger of his
being defeated. The few witnesses who swore to his presence at
the Musters had acquitted themselves badly and four of those
questioned had denied his presence. Worse still, these four were
included in the long list of witnesses waiting to appear for the
Defence.

He had managed to rake together a few of the latest arrivals
in the Regiment, including Colonel Prior, to testify that his
manner towards his officers had been on the whole conciliatory.
But the volume of evidence waiting to be given, unless rebutted
in some way, would be overwhelmingly against him. Behind the
scenes he was making plans to do this.

Before the Trial, Smales had handed in a list of witnesses for
his Defence, but had received no official information as to who
would be testifying against him, his request for a list of Prosecution witnesses having been turned down on the grounds that
while it was allowable for the Judge Advocate to supply one, it
could not be demanded as a right. This discrimination, by giving Crawley precise information as to the opposition he had to
contend with, clearly gave him an enormous advantage.

The Defence list included the R.S.M. and other sergeant-
majors, whose evidence, in theory at least, would strengthen the
Paymaster's case. But as Crawley well knew, these N.C.O.s

were far more vulnerable than officers. So, like any good tactician, it was at this point he decided to attack.

The first official information of his plans appears in a statement made to Captain Renshaw by his Troop Sergeant-Major, whose name was Moreton. He said he had attended a meeting at R.S.M. Lilley's house on 4 April to hear the proceedings of the first three days of the Court-Martial read out. When he arrived, 'the door and windows were all made fast', and the proceedings were read out to himself and four other sergeant-majors by Sergeant-Major Wakefield.

Then—'On Sunday evening, the 6th of April, 1862, at about half-past nine o'clock pm, I was under the verandah of D Troop talking to S.M. Wakefield and Sgt Bernard; about a quarter of an hour previous to this they were told in my presence by Sergeant-Major Lilley, that the Commanding Officer had heard we had seen the proceedings of the Court; he also said that "the old fellow", meaning the Paymaster, felt very much annoyed at the Colonel having heard of it. Sergeant-Major Lilley then went away, and the following remarks were made by Sergeant-Major Wakefield, "By God, the man who gave the information I will poison him, or in some way make away with him, for he is not fit to live." . . . They both swore that they (Wakefield and Bernard) would swear, were they ever called before the Court-Martial, that they had never seen the proceedings at all; this is all that occurred at the time.'

On receiving this statement, Crawley despatched it to General Farrell with a covering letter in which he pointed out that as witnesses were excluded from the Court except when giving their evidence, the Sergeant-Majors were infringing the rules of courts-martial. He went on:

'It is an act which I submit, will expose to the Major-General and to his Excellency how deep-laid a conspiracy against me and my authority exists in the regiment, and how the spirit of insubordination and indiscipline which it betrays is fostered and encouraged, even in the non-commissioned ranks and lower grades of the regiment by parties inimical to me.'

The fact that the proceedings were read in a 'secret and clandestine manner' behind closed doors and windows and at so late an hour, showed that the Sergeant-Majors were well aware that what they were doing was illegal.

He concluded:

'The Major-General will observe that this statement rests upon the unsupported testimony of one man, who doubtless would be declared to have stated a falsehood by all the parties concerned, according to their declared intention; but he is a man of the highest character, and I do not consider it possible that he could or would invent a story of this description, or that he would falsely accuse any man. The revelation of his name to the parties named in his statement would without doubt subject him to persecution and annoyance at their hands and might lead to the fulfilment against him of the sanguinary threat uttered by T.S.M. Wakefield. I have, therefore, kept his name a secret lest evil might befall him.'

These arguments for not revealing the identity of his informant need hardly be taken seriously. The Sergeant-Majors would probably try to deny the accusations but Crawley's apparent concern for the life of Moreton should his name be divulged was an obvious excuse for keeping open this line of secret information. That this was well understood at H.Q. Bombay is apparent from their reply:

'Sir William Mansfield remarks that at this stage further action in the matter concerned *does not seem to be expedient* and his Excellency observes that such appears to be Lieutenant-Colonel Crawley's wish.' His letter would be recorded in that sense and remain strictly confidential and 'not out-lost of'.*

Meanwhile, at Mhow, the President of the Court had issued a statement that he had been informed of the proceedings being read in the R.S.M.'s quarters and warned of the very serious consequences should the offence be repeated.

This was the first intimation Smales had that there was an informer amongst his witnesses. He still had no idea of who it might be and was to remain in ignorance for two more weeks. He did, however, lodge a protest at the manner in which the Prosecution was collecting evidence; it was rejected by the Court as objectionable and irrelevant.

On 21 April, almost a fortnight later, Smales opened his

---

* Sir William's letter was signed by his Acting Deputy Adjutant-General, Frederick Thesiger, the future Lord Chelmsford and elder brother of Captain C. W. Thesiger of the Inniskillings. This was of course another line in the 'old boy' net-work.

4

defence, and while he was examining the first of his many witnesses, the sequel to the events just described was enacted.

On 24 April Moreton went to his Troop Captain with further information which was again recorded in a written statement. He now claimed that R.S.M. Lilley had passed on to the Sergeant-Majors, including himself, papers to read on which were written part of the Paymaster's Defence and that he had taken copies of some that were, he said, in the Paymaster's own hand-writing. His statement ended:

'On Wednesday morning, 23 April 1862, at about half-past ten pm, I went to T.S.M. Wakefield's room, and in the course of conversation I said to him, "You have had these papers ever since Sunday:" he replied, "Oh yes; it does not matter, we can keep them as long as we like, it is only a rough copy." I then said, "It is a wonder the Paymaster gave you the papers after what happened about the last" (meaning the remarks of the President of the Court-Martial on account of the first three days' proceedings being read out to the Sergeant-Majors). Sergeant-Major Wakefield replied, "Oh, he does not care a d—— now that the proceedings are allowed to be published." '

On receiving this information, Crawley passed it in confidence to the President of the Court and consulted with General Farrell as to the best use to be made of it. Officially, however, he took no action until the evening of Saturday 26th, when he sent for R.S.M. Lilley, and T.S.M. Wakefield and another named Duval, to come down to his house.

There in the presence of Captains Renshaw and Curtis* and Lieutenant Davies, he examined them on the information supplied by Moreton and eventually obtained from each a written statement. In effect these were admissions to having read the proceedings of the first three days of the Court-Martial and part of the Paymaster's defence before it was delivered.

Crawley then wrote to H.Q. at Mhow explaining what he had done, although to Farrell who was now well involved in the whole affair, this could have been no surprise, and the letter was probably intended more to enlist the support of Sir William Mansfield than to inform the Mhow authorities.

He wrote:

* Curtis had recently joined and was living in Crawley's house.

'. . . It will be apparent, that the reading of the proceedings of the trial on the first three days, as previously reported by me, is an established fact, and that it was done in the clandestine manner therein described, with closed doors and windows, thereby evincing the guilty knowledge these non-commissioned officers had of the impropriety of their proceedings.

'I have the honour to report that I have placed R.S.M. Lilley and T.S.M.s Wakefield and Duval under arrest, on a charge of conspiracy against me, their commanding officer, and to request that for reasons known to the Major-General, in connection with a sanguinary threat uttered by one of the T.S.Ms., I may receive the Major-General's sanction to placing all of them in *close arrest*, under a sentry, pending a reference to the Commander-in-Chief.

'There can be no security against these men being tampered with by Paymaster Smales or his agents, nor safety for the person who has brought these matters to my notice, unless this is done.'

In support of his charges against the Sergeant-Majors, he sent two further statements by Moreton. The first described an incident on the parade ground on Wednesday evening:

'Somebody mentioned about the Court-Martial. The R.S.M. then said "I suppose we shall be called upon in a few days to give our evidences," and he turned half-round and raised his hand and pointed towards Colonel Crawley's bungalow, or, at all events in that direction, and said, "By God, if that b—— down there cross-questions me, I'll let out about his carryings on. He has often made me shed tears, and I'll pitch into him, or stick it into him stiff enough." S.M. Wakefield said, "I recollect you Sergeant-Major shedding tears once, but only once; you came and sat by me in the mess-room on a form at Nuggur." We then separated and went to our quarters.'

The second was an account of what took place after the interrogation at Crawley's bungalow when the Sergeant-Majors were going back to their quarters.

'. . . The R.S.M. asked me, "Moreton, did you let out the whole of it?" I answered, "Yes, Sergeant-Major, every word of it; I am not going to perjure myself for anybody." He (the R.S.M.) replied, "By God, I am done for then; I may commence and sell off as quickly as possible." T.S.M. Wakefield replied, "Oh no, wait until you see how things turn out, it might not be as bad as you think." T.S.M. Duval then spoke up and said, "No private conversation here, for I had a little private conversation last night and T.S.M. Moreton told every word of it to the Colonel today." I answered, "Yes, of course I did, when you denied the truth before your Commanding Officer, and you are not a man or you would acknowledge the truth." This is all that occurred.'

This account shows how specious had been Crawley's concern for Moreton's safety.

When the Court resumed on the following Monday, its members could hardly have been surprised at receiving yet another protest from Smales. The Prosecutor, he said, had summarily placed three of his witnesses, the most respectable N.C.O.s in the Regiment, in close arrest with sentries over them. He sought the Court's protection against such proceedings, adding, that if they were allowed to continue he would soon have no witnesses to produce. The Prosecutor's conduct had been most unusual in that he had constantly employed an amanuensis, a defence witness (i.e. Moreton), long after he had been warned to appear for the Prisoner.

The protest and appeal were rejected, though the Court did, in this instance, go to the trouble of explaining that they could do nothing as the matter was outside their province. They pointed out, however, that the Prisoner really had nothing to complain about since he could still call the arrested N.C.O.s to give evidence for him if he wished.

The news of the Sergeant-Majors' arrest should have alerted Sir William Mansfield to the deep waters into which he was being led by Crawley and the Mhow Authorities. But having stepped outside normal judicial procedure in his ordering of the Court-Martial he probably found it difficult at this stage to call a halt.

On 6 May he sent two letters to Farrell; the first giving his views on the legal position and the second his directions as to what best to do. In each he made assessments that were moderate and sound, but then went on to modify them by contradictory conclusions.

In the documents he had seen, he wrote, there appeared to be insufficient evidence for bringing the Sergeant-Majors to trial for conspiracy. At the same time the conduct of Lilley and the other two Sergeant-Majors might fairly be interpreted by

Colonel Crawley as conspiracy against his authority so that he
was amply justified in the measures he had taken. Lilley was
clearly unfit for the responsible position he held and should be
removed from it.

In his second letter Sir William suggested that the best plan
was for Colonel Crawley to read carefully what he had said about
the arrest of the Sergeant-Majors, and if nothing further had
transpired to strengthen his case against them, he should order
their release. He should then 'make short but very kind address
to them and point out if any of them felt hurt, more especially
the R.S.M., the right way of proceeding was to beg to be allowed
to speak to him on the subject, a request which his Excellency
is convinced they never could have to make in vain'.

Having given this eminently sensible advice Sir William proceeded to nullify it in his final remarks: 'In conclusion, I am to
request that the Sergeant-Majors concerned are not to be
released from arrest, or the enclosed letter acted upon by
Lieutenant-Colonel Crawley, till the proceedings in the trial of
Captain Smales are entirely closed, and the Court of which
Lieutenant-Colonel Payn is President adjourned.'

He could hardly have known that the outcome of the whole
Court-Martial would hinge on that sentence.

The three Sergeant-Majors were not the only Defence
witnesses to be subjected to questionable treatment by the
Prosecutor.

Pay Sergeant Bennett, the first to give evidence, spoke up
strongly in support of Smales his immediate superior. About a
fortnight later he was ordered to move out of his comfortable
rooms attached to the Pay Office, which he occupied for reasons
of convenience and security, and go to live in what was known as
the 'Patcheries'.

The reason for his removal was that on the evening of the day
he spoke for the Defence, he threw a party at his house to which
members of the Sergeants' Mess, including the R.S.M. and the
two Sergeant-Majors (who were arrested two days later), were
invited. This, to the lively imagination of Crawley appeared as
yet one more proof of the conspiracy against him. His suspicions
were further increased when he learned that several of the guests

had been granted passes excusing them from 'watch setting' and signed by the Adjutant, Lieutenant FitzSimon.

He described his reactions later by saying that this '*Soirée*' made it clear to him that it was high time this 'hotbed of sedition should be suppressed' in the interests of discipline, and that Bennett must be placed where he could be under proper surveillance like other soldiers.

The effect on Bennett and his young wife of this change for the worse and its ominous implications may well be imagined. He had dared to cross his C.O. and if he had any doubts as to what that could mean, he had before him the object lesson of the three Sergeant-Majors now in close arrest.

A week later he felt another tightening of the screws. The Pay Office was broken into and several account books were stolen. A Court of Inquiry was set up and he was called before it to give evidence.

In it he said he had locked up the Office the night before and was talking to Mr Hudson, the Apothecary, when he noticed a European soldier some distance away staring suspiciously at the Pay Office. He left his companion and walked towards the figure to investigate but whoever it was moved away towards the Camp and disappeared before he could be recognized.

This statement was confirmed by Mr Hudson. But the following day Bennett was arrested on the grounds that his failure to apprehend the mysterious watcher raised suspicions that he was himself implicated in the robbery.

That efforts were being made to silence other Defence witnesses was revealed when Smales demanded to be allowed to question before a magistrate, Private Walsh, Crawley's personal servant, who, he said, was going about the Camp threatening soldiers who had volunteered to give evidence for the Defence.

This request was refused. Instead, Walsh and three of his 'victims', Trumpet Major McEwan, his wife and Pay Sergeant Bennett, who had married their daughter, were interrogated by Crawley himself in the presence of the Bazaar Magistrate.

McEwan, an old soldier nearing retirement, stated that Walsh had called at his house and warned him they were 'a ruined family' because he had not come forward to give evidence for the Colonel. The first stroke of ill-fortune would be the loss of the Canteen of which he was in charge. He said he feared the words of Walsh were already coming true after what had happened to Sergeant Bennett, following the loss of the books from the Pay Office. At the same time he was careful to deny all knowledge of his son-in-law's affairs.

Mrs McEwan confirmed what her husband had said concerning Walsh but added she could not believe a gentleman like Colonel Crawley, who had always been kind to their family, would ever harm them.

Bennett swore that Walsh had never threatened him, which, after his recent experience, is not surprising.

Walsh for his part denied having threatened anyone or that he had been authorized by Colonel Crawley to do so. He did, however, admit giving 'cautions' to some witnesses. In his own words, 'I said, if you do not mind yourself and do your duty, Colonel Crawley, who has done you a great kindness in giving you the Canteen, can also take it away from you'.

He was not punished in any way for the crime of which this statement was an admission, a leniency that drew surprise from even the Bombay Authorities when they heard of it later. Crawley of course disclaimed all responsibility for what his servant had done and called him a blundering harmless Irishman. But whatever may have been the whole truth of the affair, he must have felt well satisfied with the results of the man's activities.

McEwan, who was on the list of Defence witnesses, did not give his evidence and it can hardly be held against either him or Bennett that from now on they seemed to have been used by Crawley to further his ends whenever the need arose.

8

The Case for the Defence

The Paymaster's Defence lasted intermittently over the five weeks from 21 April to 24 May.

Against Crawley's few witnesses who had made such a poor showing on the question of his absence from Muster Parades, Smales produced thirteen, all of whom swore not to have seen their Commanding Officer on the Parade of 1 May, 1861, or that of 1 January, 1862. They included Major Swindley who was acting Paymaster on 1 May, Captain Weir who commanded that Parade, the Adjutant, the Pay Sergeant, the Riding Master, the Quartermaster, the R.S.M. and four Troop Sergeant-Majors.

Their combined testimony appears so conclusive that it is surprising that Crawley decided to contest the matter, particularly as his absence amounted to a purely technical offence, a point actually conceded by one of the Defence witnesses. There was no doubt either that he had expressed regret that duelling was no longer in use, indeed he admitted as much:

'I remarked that when I first entered the service, quarrels were settled in a very different way than by appealing to the commanding officer. I informed the disputants that in those days officers settled matters between themselves when they quarrelled. I said I was almost tempted to wish that those days were still in existence. . . . Such was the substance of my remarks to my officers. I have no recollection of making use of the word duelling, though doubtless my remarks pointed that way.'

As on numerous occasions it is obvious he was 'coming the old soldier' to impress the younger men. But in the light of his admission, the Paymaster's accusation could hardly be described as false, particularly as Crawley was at great pains to stress the undisciplined state of the officers. His reference to duelling could reasonably be interpreted as not calculated to promote 'concord and forbearance'.

The two quarrels said to have provoked his remarks were trivial in the extreme. The first involved Riding Master Malone, VC, and Lieutenant Revell, whose proneness to disputation had earned him a trip to China. Malone appears perhaps unduly anxious to maintain his dignity.

This is how he described the brush:

'I met Mr Revell in Mr FitzSimon's compound; he said to me, "Oh, you're riding that old screw I see."

'I replied, "He's not a screw, and I did not ask your opinion; I beg you won't volunteer any more on the subject."

'He began to say something more, when I said, "I don't wish to argue with you."

'He replied, "You're not worth arguing with."

'On this I told him I considered his remarks offensive, and that I would refer it to Colonel Crawley. All this was intended as chaff to me, but as I said before, I should never allow him (Lieutenant Revell) to make free with me in any way.'

The disputants duly arrived before Crawley, when, according to Malone, Revell was told that had he been in the Service when he (Crawley) first entered it, he would have been shot through the heart or the head. This statement was corroborated by Weir (Malone's father-in-law) who happened to be present. Revell's version of the incident is not recorded as neither side called him as a witness!

The second quarrel, between Swindley and Poett, the Veterinary Surgeon, took place on the parade ground during the November Inspection by General Hale. It was over a horse which the vet announced in loud tones ought not to have been on parade. Swindley retorted that Poett wouldn't know a lame horse if he saw one. The vet then rushed to lay a complaint before Crawley.

Quite conceivably the incident was shrewdly contrived by Crawley to show up Swindley at his irascible worst before the Inspecting General. Certainly it was the method he used to good effect in the future in dealing with this opponent.

Several witnesses swore that in settling the dispute Crawley again referred to duelling. But this was denied by Poett who seems to have thrown in his lot with Crawley after falling out with most of the other officers in the regiment.

It appeared that he had at one time been as upset at Crawley's manner as anyone. A cornet named Snell, an astute young man who managed to derive a deal of amusement from the feuding of his elders, said that Poett complained to him of Crawley's addressing him 'like a dog' and that this seemed particularly to have upset him because some Governor or other had once called him (Poett) 'the Saviour of India'.

Still on Crawley's manner, Malone recalled an occasion in the Orderly Room when Weir felt so humiliated at the trouncing he had just received that he *begged* Crawley not to speak to him again in that way before junior officers.

Major Swindley, taken to task for his slow progress in completing his horse-lines, said he told Crawley he had done them to the utmost of his ability. Crawley replied, 'Your utmost is d——d little.'

'I had been upwards of fifteen years in the Army,' Swindley said, 'and I had never been spoken to as he addressed me. My feelings were wounded and, being so spoken to in the presence of my Troop, my position was humiliated.'

Then again: 'I told Colonel Crawley that my manners and tone were taken from him; that it was hardly to be expected that I could meet him in any other way when he came to me scowling and addressing me in a most offensive manner.'

It came out in cross-examination that these two had met once or twice in England where the seeds of animosity may have been sown. The older man's inability to impose his authority reveals itself in his repeated references to age and long service, as:

'Did I not tell you on that occasion that I recollected you a little boy at Maidstone, two or three years before; and that it did not become you to set yourself up against me?'

The same lack of tact is evident in his efforts to change the regimental customs even allowing his intentions to have been good and the changes for the better. Witness followed witness to complain of his attitude.

On being told it was the established practice to keep the

grooming bags under the men's beds, Crawley had expressed incredulity. Weir introduced a note of levity by saying that he remembered a certain general in Ireland, thirty years before, who was also surprised when he heard of it. Crawley not only took him to task for his frivolity but said he could not believe the custom had been sanctioned by his predecessors and would write to Colonel Shute to find the truth.

On another occasion he called the Troop Captains to the grain shed and asked them why they did not see to it the grain was properly served out. One tried to tell him it was the custom for the grain orderlies to see it done. He exploded—'Some of you Captains have stuck up your backs against me, but by God I'll straighten them.'

A lieutenant named Stevenson said that once, when he was orderly officer, Crawley became very excited because things were not to his liking. '. . . he complained that the proper number of horses was not sent to the riding school, and also that the line sentries were not properly placed. I do not recollect the exact words Colonel Crawley made use of, but he cursed and swore in the most extraordinary manner, and said if he had the power, he would try all the troop sergeant-majors by one court-martial and break them, or words to that effect. He appeared to be in a most excited state; I could not account for it, as I saw no sufficient reason for his using such extraordinary language.'

Solemnly questioned by the Court as to how Crawley expressed himself, Stevenson said: 'He interlarded his language with oaths; he did not curse any person in particular; he was addressing himself to the whole regiment.' These extracts, representative of the Paymaster's long Defence, show that on the evidence before the Court, the case against him must have failed. But Crawley's real strength lay in factors quite outside the Prisoner's control: the determination of his superiors to support him at all costs, the whole-hearted co-operation of the Court in finding Smales guilty, and his own capacity for intrigue together with his power as commanding officer over the witnesses appearing against him.

It was clearly impossible to refute the mass of hostile evidence slowly accumulating against him; the alternative was to discredit it by attacking the characters and reputations of those who were giving it. And to achieve this there were certain priorities, for among the witnesses were some whose testimony would be far more damaging than others. At the top of the list were obviously the Adjutant and the R.S.M. who from their daily contact with him and the confidential nature of their duties could be expected to know the details of what had been going on over the past months.

In the matter of his presence or absence from parades, for instance, it would tax the credulity of any military man to disbelieve their agreed testimony, unless there was reason to believe they were perjuring themselves. Recognizing this, Crawley had already taken steps to discredit the R.S.M. He now set about performing the same operation on the Adjutant.

Lieutenant FitzSimon had been appointed to the Adjutancy not long before Crawley arrived, having defeated Lieutenant Davies for the post. An Irishman from Dublin, he had received his early military training in Austria and was 26 when he entered the British Army as a cornet in 1858. According to his record he was good at languages and passing exams but not much of a horseman by cavalry standards. He had little or no private means so the adjutancy with its emolument was important to him (as it would have been to Davies).

His value to the Defence was two-fold. From having been regularly in the company of his C.O. both in the Orderly Room and about the station, he could corroborate the evidence of other witnesses; he also knew and had been a party to some of the back-stage manœuvres against the Paymaster.

At the same time his role of Defence witness placed him in an invidious position, for higher authority was bound to take the view he had acted disloyally. Moreover he had been implicated, however unwillingly, in some of Crawley's intrigues and this would make him particularly vulnerable under cross-examination.

He swore that Crawley had sent Colonel Prior to Smales to try and get him to withdraw his letter but with the proviso that no arrangements were to be made without his approval. And probably this was true. But in the face of Crawley's denial and with Prior prudently refusing to become further involved, the Court naturally chose to ignore that part of his evidence.

In cross-examination he was accused of associating with Smales when forbidden to do so. He denied this, saying that, just before the Paymaster was arrested, he was ordered to go to his house with an official communication. He had protested he would rather send it with a covering letter. But Crawley insisted he should go personally as he might 'find out something'.

FitzSimon concluded: 'I did go to the Paymaster and gave him the official message, and requested him on that occasion not to speak to me on any of these matters, as I should be obliged to tell the Commanding Officer of them if he did, and I wished to keep clear of the entire business myself.'

The inconsistency of this assertion of a desire to remain neutral with his appearing for the Defence was unlikely to be lost on the Court or on those who would later review the proceedings.

He said he had been required several times to write out statements on conversations between Crawley and Smales. Describing what happened when the paymaster was refused leave, he said he had been given a list of questions to answer while Crawley stood over him.

He went on: 'I gave answers to the questions as much as possible according to my own idea, but there were several words the meaning of which I did not like; but still, when they were explained by Colonel Crawley, I had to write them down. I supposed Colonel Crawley understood them better than I did.'

Crawley: 'Have I not on all occasions when asking you questions, told you I wanted you to say nothing that you could not swear to before a court-martial?'

FitzSimon: 'Yes, I think Colonel Crawley did say some words to that effect, once or twice, when I remonstrated about writing down what he wanted me to do, but at the same time he pressed me with the same words.'

The Adjutant's defection undoubtedly strengthened the Paymaster's case, but for Crawley it was not the catastrophe it first appeared. For while FitzSimon could give away much that he wanted to remain hidden. Crawley knew a good deal about the past behaviour as well as the weaknesses of his Adjutant and was well able to exploit them to his own advantage. It may perhaps be too much to suggest that he had yet marked out the role FitzSimon was to play but his astute mind must have already realized that here was one of Nature's 'fall guys'.

By contrast with the state of mental subservience to which FitzSimon had been reduced, the reaction of Quartermaster Wooden when he was called upon to write out a statement was markedly different. For when Crawley sent him his version of the conversation 'as a guide', Wooden returned it saying he would rely on his own memory.

Like Renshaw he had at one time been engaged in following Smales about in an attempt to 'catch him out', but had eventually become tired or disgusted with playing secret agent for Crawley. In his evidence he said that Smales had asked him if he thought it honourable for one officer to make secret reports on the conversation of another. He admitted it was not, but that he had no choice because his Commanding Officer had ordered him to make them.

On 7 May, the twenty-second day of the Trial, R.S.M. Lilley was marched under escort into the Court Room to give his evidence. Several features combined to make the occasion of peculiar interest.

He had been in close arrest for eleven days. His rank, even then, was one of great importance, entitling him to respect. He seems to have been popular throughout the Regiment and was at the time the object of sympathy, not only from the circumstances of his arrest but because of tragedies in his domestic life; for his two young children had died recently and his young wife was now in the last stages of consumption.

But while these were the factors colouring the views of the majority of the European inhabitants of Mhow, they were of far less significance in the eyes of the small group of officers in charge of events. To them Lilley was an N.C.O. and as such could expect none of the protection and consideration afforded an officer of even the most junior rank. The Court would listen to what he had to say, but they already knew he had been accused of conspiracy by his C.O. and that was enough. For all the effect his words were likely to have on the course of the Trial he need hardly have put in an appearance.

Questioned by Smales, he said he was on the Muster Parades of 1 May, 1861 and 1 January, 1862, and did not see Colonel

Crawley at any time while the musters were being taken. Between the dates mentioned he had no recollection of seeing his C.O. on any Muster Parade except two, those of 1 November and 1 December. He attended the C.O. on all monthly parades and if Colonel Crawley was present he would know it.

His evidence on Crawley's harshness is at times surprisingly unconventional.

'Have you ever heard the Prosecutor compare the 6th Dragoons to yeomanry or a mob when on parade?'

'Several times.'

'Have you heard Lieutenant Colonel Crawley speak in disparaging terms of the system of the regiment; if you have state the remarks as near as you can recall?'

'I have heard him say "You may do for Colonel Shute, but you won't do for me; your system is rotten and you can't ride."'

'Has the Prosecutor's language and manner towards you when on parade or duty been more harsh and unusual than former commanding Officers?'

'Yes.'

'Were you distressed to tears by the Prosecutor's language towards you?'

'Yes.'

'Can you state to the Court any instance of this harsh language by which you were so much distressed?'

'Yes I can; the first instance was at Nuggur, in Colonel Crawley's house, where he held his office. I was ordered out of his office, I may say, like a dog, by his saying, "Go away, Sir," in a very harsh manner, several times; and when I was leaving the house, when I had gone several yards, I could hear the same sort of language, I supposed towards me. The adjutant was present at the time. . . .'

'On another occasion at Nuggur, some orders had been published by Colonel Crawley; his orders were not explicit, the troops did not understand them, and the parade was not according to the Colonel's wishes. I gave out the order. It was for riding drill. The next morning after the riding drill was dismissed he called me over in a very harsh manner, "Come here, Sir. What is the meaning that my orders were not carried out?" I was about to make a reply, when he spoke in a very harsh manner, "Silence, Sir, hold your noise." That morning I felt very much affected at the manner I had been spoken to. . . .'

'There was another instance in Mhow, in the presence of the adjutant; I was told by Colonel Crawley that he would have me up night and day if he thought that proper. There were several other instances, but I could not name them all. . . .'

'I heard Colonel Crawley speak to the officers commanding troops on one occasion at Nuggur when they were assembled one evening at the head of the lines. When the officers assembled I used to go a respectable distance from the officers. On this occasion I was about 40 yards off. Colonel Crawley spoke in a very loud harsh manner to the officers by saying, "Gentlemen I would have as little compunction in sending you to morning and evening stables as I have of going to dinner." I never heard a commanding officer speak to officers in so harsh a way before. . . .'

'I have served in the 6th Dragoons 18 years and between 4 and 5 months. I shall have been R.S.M. for 7 years on 3 July, 1862, to the best of my belief. . . .'

'In Kirkee (I do not recall the date), Colonel Shute told me he had recommended me for the adjutancy of the regiment. He said he was very sorry to say he had received an answer that it would not be given in the regiment at the time, but that I was noted to the Horse Guards when an appointment would offer.'

Smales then asked Lilley to state the most recent harsh and unusual act against him by the Prosecutor.

His reply was: *'The last act is at the present time, by a sentry being placed at my bed-room door, where my wife is lying. The door is quite open; the sentry is posted about two feet from my bed.'*

Here the Prosecutor handed in a paper to the Court. 'Mr President and Gentlemen, with your permission, I wish to put a question to this witness which is not immediately before the court, but which it is absolutely necessary for me to put with regard to his reliability, as I have reason to believe he has been tampered with. . . .'

The question was whether R.S.M. Lilley had read the proceedings of the Court in defiance of the orders of the President, and this, after consideration, the Court decided should be put.

Lilley replied that he had read part of the Defence on the Sunday evening before it was delivered in Court or published in the Poona and Bombay papers. He said he had received the papers from T.S.M. Wakefield and not from a native as he had previously told Colonel Crawley. (The implication here seems that Lilley had lied at first so as not to involve Wakefield.)

When the R.S.M. had concluded his evidence and been escorted out of the Court-room, Smales handed in two protests. The Prosecutor, he claimed, had examined the Defence witnesses, not to disprove their evidence but to throw reflections on their characters. Not only had R.S.M. Lilley's character been impugned but he had been asked whether he had read the Court's proceedings, a question which, as Smales pointed out, *he* had wanted to ask the Adjutant but had been prevented by the Court.

After consideration of the protests, the President momentarily at least, seems to have realized there had to be some limits to the Court's partiality and told the Prisoner he could now put his question.

FitzSimon's reply revealed that Crawley had himself committed the very offence for which the Sergeant Majors were being held in close arrest. That is he had read out two or three days' proceedings to the Adjutant *to make sure that the evidence he would give would be consistent with what had gone before.*

9

Sudden death

When the Paymaster closed his Defence on 24 May, the three Sergeant-Majors had been in close arrest for a month, during which time, except for a brief appearance in Court and two short periods of exercise each day, had been confined to their quarters.

It need hardly be said that keeping them in confinement without being charged was a direct violation of Army Law. Yet the whole proceeding had not only been condoned but ordered by the C.-in-C. of the Bombay Army, so it may be assumed such arbitrary imprisonment was not uncommon in India at the time.

The official documents relating to it would eventually appear before the Authorities at the War Office and the Horse Guards, so that Sir William Mansfield must have expected in the normal course of events that any plausible explanation he could produce would be accepted in those quarters without question. If this was the case the General may be accounted unfortunate. He had taken short cuts in the hope of settling the continuing problem of the disputes between Crawley and his officers; he could hardly be expected to foresee the event that would inflate the internal squabbles of a regiment in an obscure Indian outstation into a great national issue involving the whole British Nation.

That event may be told as briefly as when the news of it struck the small community at Mhow on the morning of 25 May, 1862—*R.S.M. Lilley had died suddenly in the night.*

The varied emotions aroused throughout the Regiment at the unexpected death of so important a person can well be imagined, as can the fact that, from the abnormal circumstances attending the death, they were far more deeply felt and lasting than was usual in an age when sudden death was a commonplace. But here the first concern is with the reactions of those who knew themselves to be closely involved in the tragedy.

Turnbull, the Senior Medical Officer, who had not himself attended Lilley, carried out a post-mortem and diagnosed the cause of death as apoplexy. His Report included the following:

'He was a strong healthy man, inclined to corpulence; about 17 or 18 stone weight. Had always enjoyed good health; was imprisoned without trial 26 April, reported sick when in close arrest on 24 May on which day he died.

'The excessive heat at this season of the year; the constitutional predisposition of the deceased to congestion; the peculiar and painful circumstances of his position; the serious illness of his wife causing depression of spirits, with bilious and nervous derangement, induced by a sedentary life attendant on close arrest, in a man of the Sergeant-Major's active habits, probably acted as exciting causes to produce the complaint from which he died.'

In this the implications regarding the responsibility for Lilley's incarceration are all too clear, which partly explains why the Mhow Authorities delayed reporting his death to H.Q. at Bombay. The first Sir William Mansfield heard of it was through the Indian papers and what he read, not surprisingly, roused his apprehension. On 2 June he sent the following telegram to General Farrell at Mhow.

'It is said in the *Times of India* that the Sergeant-Majors of the 6th Inniskillings are subjected to the severest confinement, amounting to punishment. The Commander-in-Chief, while disbelieving this, begs you to report all the circumstances connected with them. An arrest should never be of a penal character. On the Court-Martial being adjourned, instructions about the men should have been asked.'

The same day he despatched a letter with an enclosed Memorandum. In the first he pointed out that when he had expressed his opinion that the Sergeant-Majors should not be released from arrest until the end of the Court-Martial, he fully expected this to occur two or three days after his letter arrived. He now learned from the newspapers that the proceedings had been adjourned indefinitely because of the sickness of Paymaster

Smales and that the men were still in close arrest. It might have been inferred, he said, that this would be quite contrary to his wishes.

He asked General Farrell to proceed personally to the Orderly Room of the 6th Dragoons and there get his Adjutant General to read out to the assembled Officers and N.C.O.s, first his letter of 6 May and then the enclosed Memorandum. He would feel obliged if General Farrell would also himself address a few appropriate words to the N.C.O.s.

Sir William could hardly evade the responsibility for ordering the arrest of the Sergeant-Majors. But in his Memorandum he attempted to justify what he had done and shed some of the responsibility on to Farrell for the way his orders had been carried out. This part of the document was to be of the greatest importance. It read:

'. . . . His Excellency calculated that that letter (6 May) would reach Mhow about 12 May, and that the proceedings of the Court-Martial, before which Captain Smales was being tried, would have been closed on or about that time. It appeared to His Excellency necessary for the safety of the Sergeant-Majors under arrest, *that they should not be exposed to further temptation*, as it seemed to Sir William Mansfield *that they had most narrowly escaped trial on a charge of conspiracy* against their Commanding Officer. Besides which it was absolutely necessary in the interest of discipline and justice, *that the charge of conspiracy should not be condoned* until the investigation proceeding in the trial of Captain Smales had shown whether or not there were more facts which might be quoted in support of such a charge. His Excellency accordingly, in another letter, desired that they should not be released until the proceedings of the General Court-Martial were closed. He had been informed that the Court-Martial had been adjourned and that the Sergeant-Majors were still under arrest. This was quite contrary to his intentions and further instructions regarding them should have been sought.

He had learned with much sorrow that Sergeant-Major Lilley had died, but of this he had received no official report. On 6 June Major-General Farrell carried out the instructions from his superior including the delivery of a few well-chosen words to the N.C.O.s, as follows:

'Sergeants Wakefield and Duval, you are now released from arrest and from the consequences of your folly. Circumstances over which I had no control have lengthened the period of your arrest and confinement, and by the will of the Almighty, your comrade, Sergeant-Major Lilley, has been taken from among you and I deeply deplore his sudden death, and that he cannot have the consolation of forgiveness granted to him in your presence.

'All of you must bear this in mind and remember it to the last day of your service, that subordination to the authorities placed over you is the golden rule of conduct for all soldiers serving Her Gracious Majesty. The late transactions, so clandestine and so improper, so closely resembling mutiny towards your commander, have brought all this trouble and sorrow upon you; the consequences of such conduct were inevitable, and I trust you will hereafter show by your conduct that you have seen your fault in its true light and strive to regain the confidence of your commanding officer.

'To the rest of you I would say that discipline and your own honour and happiness can only be obtained in right feeling and subordination.'

In his report to Bombay, the author of this effusion wrote how pleased he had been with the respectful and cheerful demeanour of the N.C.O.s, though whether he accurately represented the feelings of his captive audience is hard to say. He no doubt wanted to allay the apprehensions of his superior, and may have conveniently overlooked their ability to dissemble in a situation of that sort.

The long interval between the Closing Speech for the Defence and the Prosecutor's Reply (two weeks) had been caused, not as Sir William Mansfield was led to believe, by Smales' sickness, but by another death on the station—that of Mrs Taylor, Crawley's mother-in-law, whose end was not sudden and who must have been greatly saddened by all the strife that surrounded her.

Her death goes some way to explaining why Crawley delayed ten days before sending in his report on Lilley's. But apart from this he was evidently anxious to present his part in the events leading up to it in as favourable a light as possible. He gave two reasons why the sentries had been placed inside the rooms with the arrested Sergeant-Majors. The first was to avoid their being exposed to the sun and hot winds; the second to prevent the prisoners being 'tampered with'. There was, he wrote, no reason why the sentry's presence should have interfered with the domestic arrangements of the Lilleys, for be believed there were three or four rooms in their quarters. He had learned a few days after the R.S.M. had been arrested that the sentry had been most injudiciously placed by Lieutenant FitzSimon so as to give cause of annoyance to the Sergeant-Major's wife.* On hearing of this he 'immediately sent Mr FitzSimon himself to withdraw the sentry from the presence of Mrs Lilley and to place him where he could do his duty with equally good effect without giving annoyance to her'. Then, as soon as the Paymaster closed his Defence on 24 May, he had sent a letter to General Farrell asking for permission to change the prisoners from close to open arrest. But by the time this was received on the following day, R.S.M. Lilley had unfortunately died of a sudden attack of apoplexy at 4 o'clock that morning.

The proof that Crawley displayed such sudden concern for the welfare of the Sergeant-Majors is slender. General Farrell, whose support for his subordinate remained unwavering, gave a different version of the application to have the nature of the arrest altered. He said that Crawley sent him a *note* on the afternoon of 24 May and an official letter on the morning of the next day. The note, however, mysteriously disappeared and only turned up several months later, and the letter was, of course, written after Lilley was dead.

Having shifted the responsibility for posting the sentries on to FitzSimon and demonstrated his humanitarianism by asking for the release of the prisoners as soon as it was 'safe' to do so, Crawley must have decided that some concrete evidence was needed to counteract the unfavourable impression created by Turnbull's Medical Reports on the dead man. His requirements were met on Sunday, 7 June, a fortnight after Lilley's death.

Lieutenant Davies, whose recent (and predictable) appointment as Acting Adjutant had greatly strengthened Crawley's hand, produced a bill sent in by a Parsee sutler attached to the

* It was actually eleven days.

Regiment, containing a long list of intoxicating drinks said to have been delivered to the R.S.M.'s house over the period of his arrest. It showed no less than 23 bottles of brandy, 12 pints of ale, 2 bottles of Port wine and 1 bottle of gin.

All this, wrote Crawley in his Report 'the unfortunate and misguided deceased must have drunk'. It also showed clearly that, before his arrest, the R.S.M. must have been a heavy drinker and, 'about the time he was engaged in concocting and aiding in his conspiracy against me, viz. between 6 and 26 April, he was liberally supplied with the maddening drink which caused his lapse from the path of duty and eventually his sudden death.'

'It will not do,' he went on, 'for Surgeon Turnbull to ignore among the causes which he assigns for that event, the influence of such an amount of fiery stimulent upon the frame of a man violently predisposed, according to his own account, to an attack of the disease which caused his death'.

He added that there was good reason to believe that Sergeant-Major Wakefield's illness was due to the same cause but he could state nothing tangible since he had no 'positive proof' as in the case of Lilley.

The morning after he received the bill, Crawley sent for Turnbull and in the presence of Davies told him of the large quantities of drink that had been delivered to Lilley's quarters, all of which, he said, must have been consumed by the R.S.M. For only Mrs Lilley, who was a sick woman, and a native servant had been allowed into the house, nor was any liquor found in the house after his death. In the light of these 'facts' he invited the Surgeon to reconsider his original Report.

Turnbull replied that he knew nothing of Lilley's having drunk spirits during his arrest and since no one had previously suggested that he had, there was no reason to make inquiries on the matter. There was, therefore, no reason why he should add to or amend his Report. He was warned of the grave responsibilities he would incur by ignoring the new evidence but remained adamant and left.

Two hours later he returned and handed in the following addenda to his Report:

'In addition to the above, I wish to add that it has been brought to my notice that the deceased was in the habit of drinking a considerable quantity of brandy during the period of his arrest, and, on inquiry, I find the above to be correct. It is my opinion that this, in connection with the other exciting causes stated, was calculated to increase the predisposition to an apoplectic seizure, from which he died. The universal reputation which the deceased enjoyed in the regiment for sobriety and steadiness, and the Medical Officer in his daily visits never having noticed any symptoms of intemperance, precluded all suspicion of his being addicted to the use of spirits.'

Now whether Crawley considered that, from its guarded nature, this addition would not strengthen his case or for some other reason, he decided not to forward it with his letter to H.Q. but returned it to Turnbull with the comment that it had arrived too late for the post and that in view of his own detailed report on the subject, it was unnecessary.

10

The end of the Trial

The Court-Martial was not concluded until 7 June, by which time the proceedings had dragged on for almost ten weeks. The Closing Speeches were also of great length, though here it will suffice to indicate those parts most closely connected with future events.

The Prisoner delivered his on 24 May, the day before Lilley's unexpected death. In it he stressed the many difficulties he had encountered in putting forward his case and contrasted them with the assistance the Prosecutor had received in preparing and presenting his. Because the normal court-of-inquiry had been dispensed with before he was charged, he had been prevented from producing much of the evidence he required. The Prosecutor, on the other hand, had been allowed to introduce all kinds of irrelevant matter—to libel a whole Regiment and its previous Commanders, to rake up old quarrels and to drag in his difficulties over horse-lines, Mess-dinners and the like—all of which had nothing to do with him or what he was charged with.

The Prosecutor had accused him of tampering with witnesses, yet had himself arrested men who were to appear for the Defence and had confined them for weeks without trial. At the same time his servant had gone about the camp intimidating others and preventing them from appearing in Court. At the start of the Trial the Prosecutor had been so short of witnesses that he ordered the Adjutant to canvas the Regiment in an attempt to find men willing to swear they had seen him on parade. Yet of all those he interviewed, he managed to muster only two sergeants and two privates.

Smales remarked particularly on the evidence of Lieutenant Bennitt, which, he said, could well have proved decisive had it not by good fortune been disproved. Had any Defence witness dared to give evidence of that sort, who could doubt he would immediately have been charged with perjury and his fate sealed.

He concluded:

'Seldom, if ever, do I think it possible that any officer could have entered upon and gone through such a protracted and painful trial as that which has fallen to my lot. Unaided and alone have I sat, day by day, week after week, without the remotest indication of a shadow even of sympathy for my position . . . while the Prosecutor enjoyed the ability and experience of the Deputy Judge Advocate, who, from assisting him in the preparation of the charge, has gone, as it were, shoulder to shoulder with the Prosecutor in order to secure a conviction. . . .

'All these advantages I grant the Prosecutor willingly, knowing as I do, and having the fullest assurance that every member of the Court will judge me, not by the animus of the Prosecutor, but the evidence of the witnesses, the true and sterling merits of the truth elicited, and that the charge and nothing but the charge, will be brought to bear in their minds when their verdict is decided upon. I therefore, with the greatest possible confidence, retire from my defence, leaving myself confidently in their hands; with every possible respect, I tender to them my grateful acknowledgements.'

Whether, in view of the treatment he had received, Smales had any real faith in the Court's impartiality is improbable. At all events, because of the delay in the Prosecution's Reply, he had to wait sixteen days to find if it was justified.

This began with a review by the Prosecutor of the scanty evidence as to his presence on the Muster Parades, the flaws in which he blamed on the hostility of the Defence witnesses. He then proceeded systematically to attack their characters. That of the dead R.S.M. he dismissed out of hand on the grounds that he had engaged in a mutinous plot against his authority. Of the Adjutant he said: 'As every person acquainted with Lieutenant FitzSimon knows that he is so blind that he cannot distinguish any person at five yards distance, so as to be sure of his identity, I leave the Court to judge of what value is his evidence as to my being on parade or not.'

That a strong animosity existed between himself and Swindley has already been shown. But it was important in view of this opponent's rank that he should be further discredited. To this end he gave his version of an incident that occurred when he first joined the Regiment at Nuggur in which Swindley was (more or less) involved.

When he and his party arrived from Poona at 9.30 am after an exhausting train journey of 17 hours, he sent for the mess orderly and told him to send breakfast to his house. After waiting some time the man failed to return or the breakfast to appear so he despatched his butler to find the reason. He then learned that the meal was ready but that it could not be brought as no eating utensils were allowed out of the mess. Only after his butler had borrowed some from the house of an officer nearby were they able to sit down to a meal.* It was then 11 o'clock and the Court would understand how hungry they all must have been.

He was later told by the messman that Captain Swindley, the Mess President, had refused to allow the articles required out of the mess because of a rule to that effect.

Another topic on which Crawley elaborated his case against Swindley was a dinner given soon after he arrived, in honour of Brigadier Hobson who was inspecting the Regiment for the second time in a few weeks. He now alleged that some of the officers, including Swindley, had refused to pay for the special wines for the guests on the pretext that they had just given one dinner for the Brigadier and his staff and they regarded the second inspection as a punishment. As a result of their parsimony he had been obliged to pay for these wines out of his own pocket. He could not, however, produce any proof of this because Major Swindley had since destroyed the wine books in order to remove the evidence of 'a transaction so discreditable to himself'.

Crawley had made no reference to this matter during the Trial, though he had cross-examined Swindley on a conversation

---

* The officer who played Samaritan was almost certainly Smales. Ironically Crawley could hardly name him, nor could Smales claim credit without confirming the story.

between them over the issuing of invitations to the dinner. He now added the following to what had already been said.

'When he (Swindley) said, "I never forgive", as detailed in his evidence, I replied, "I am sorry for that; I thought I saw you at church last night?"

'He said, "Yes, I was on duty there with the troops, but I never go there unless I am obliged." '

Then later, warming to his theme:

'Well might Major Swindley say, when I put him on the rack of cross-examination, as reported in the account published in the newspapers,* "This is very unpleasant; I cannot but feel that I am now being placed upon my trial." Assuredly so, Major Swindley! you *were* being placed upon your trial. I have proved from your own lips that you never forgive, that you are of a sordid and malignant disposition, and that you are insubordinate, disobedient, insolent and defiant to those placed in authority over you, that you are tyrannical and abusive to those under you, that you are arrogant and insolent among your associates, for you are the man Lt. FitzSimon told me, shortly after I joined, made the mess of the regiment so disagreeable by your overbearing manner and conduct. . . .'

The third officer subjected to attack was Captain Weir and here too Crawley's assertions may not have been too wide of the mark. He said that many troop captains (Weir in particular), and sergeant-majors were dissatisfied with him because he insisted they should buy a proper supply of line equipment for their horses. To do so, they complained, would seriously affect their profits on their allowance. In that case, he had told them, they must have mismanaged their affairs for he knew the captains of his old Regiment always reckoned on a profit of 100 rupees a month which they shared with the quartermaster.

He also brought up the insubordinate letter sent by Weir following the adverse report made by Brigadier Hobson. Weir had apparently 'boasted' of having been reprimanded by the C.-in-C. but not ordered to apologize or withdraw it. Crawley added: 'In drawing attention to all this I have no wish to have it inferred that the letter alluded to was the production of Captain Weir. He could not have penned a sentence of it, but it was and

* Not all remarks at the Trial were officially recorded. The Court, for reasons that must be suspect, excluded some.

is perfectly well known by whom the letter was written, nay, the writer of it boasted to me that he had written it;* poor Captain Weir paid the penalty of adopting as his own and placing his name to a letter characterized as above.'

The implied criticism of his predecessor's running of the Regiment made throughout the Trial obviously placed Crawley in a dilemma. Towards the end of his speech he tried to play this down. He also harped on the recurring theme—the great difference in discipline and conduct of officers and men.

'I solemnly deny that I have ever reflected upon Colonel Shute in any way whatever. The state in which I found the regiment, the setting up of the men, their steadiness on foot parade, their orderly and respectful demeanor to their officers, the small amount of crime, and the general good feeling existing among the non-commissioned officers and soldiers showed me how good and beneficent had been his command of the regiment, so far as the soldiery were concerned . . . but it was quite another story as regards the conduct of certain officers, who showed me from the onset that he had been unsuccessful in his rule over them; for, as the prisoner himself put it, they were in a state of disorganization—of chronic insubordination.'†

In conclusion he wished merely to inform the Court of the position he found himself in when he assumed command. He had come 'red hot' from one of the finest cavalry regiments in the British Service in the strictest garrison in Great Britain. Was it surprising then that he should view with a critical eye everything connected with his new corps.

From the day he ventured to disagree with the Prisoner and his adherents his life had been occupied in demolishing the fabrications built up against him. 'Mind and body must give way under such pressure, and the heavy affliction under which I and mine have lately suffered and are now suffering has been mainly brought about by the worry and annoyance wrought by the prisoner. His upas‡-like shadow has fallen on my threshold, and

---

* Quartermaster Wooden; said to be the best letter writer in the Regiment.
† The assertion that Smales had used the words quoted was of course improper. He had in fact said that *Crawley* was trying to prove the Regiment was disorganized and in a state of chronic insubordination. Quoting out of context was one of Crawley's favourite ploys.
‡ A fabulous Javanese tree, so poisonous as to destroy all life beneath its branches.

has converted my heretofore happy home into a scene of mourning and woe! Alas! mine is not the only one.'

He had done his duty in laying before the Court all the intrigues against him. He was no lawyer and he had received no assistance from any. He was but a plain soldier with thirty years service to the Queen (man and boy), with an unblemished and stainless reputation.

He had had scarcely a day to himself in the last fortnight due to the prisoner and his plots and he hoped the Court would make allowance for any failure of his to reply to all the details of the prisoner's voluminous defence.

He would now leave the matter in their hands, confident that he would receive 'ample and strict justice' and that 'a stern retribution' awaited the wrongdoer.

Two days later the Court delivered its verdict.

It found the Prisoner guilty on all three charges and sentenced him to be cashiered. It had this to say about some of the main Defence witnesses:

'The Court considers it would fail in its duty did it not remark on the strong animus displayed in giving their testimony on this trial by Major Swindley and Captain Weir; on the evasive character of the evidence of Lieutenant and Adjutant FitzSimon; and more particularly on the reprehensible conduct of Captain Weir in not having made his subaltern, Lieutenant Bennitt, aware that he had been on a committee at the time he supposed himself to have been on the muster parade on the 1st May 1861 of which fact he (Captain Weir) was fully cognizant previous to Lieutenant Bennitt's appearance in Court.'

11

The departure of Mr Smales

Before his trial Smales had been convinced that Crawley intended to ruin him professionally as the surest way of removing him from the Regiment. There had been probing into his management of the Band's Fund in October, his arrest for refusing to make payments to Renshaw out of office hours, and finally his being refused leave until a particular account was settled, were all intended to reflect adversely on his ability and integrity as a Paymaster.

To the extent that the charges against him at the Court-Martial were not connected with his work but opened up much wider issues, his counter-attack on Crawley in his insubordinate letter had been successful. But only momentarily. Throughout the long trial, Crawley had never lost sight of his main objective, that of proving his opponent unfit for the responsible post he occupied. Indeed, as the proceedings dragged on the need to do so became more urgent, for he knew that even when the Court had found in his favour, that would by no means be the end of the matter.

Experience had taught him that Smales was not the man to accept defeat without protest. And while the authorities in India could be relied on to ignore or nullify his voluble complaints, the idea of Mr. Smales hammering on the doors of the Horse Guards and the War Office and calling loudly for justice was a disturbing prospect.

Fortunately for Crawley, events during the trial had greatly increased his power over his subordinates in the Regiment. And at the same time, Smales, who had always been vulnerable by reason of his position as cashier, was preoccupied with defending himself. He was also limited by the conditions of his arrest from guarding against attacks on his professional integrity.

And though Crawley was also involved in the Court proceedings, in most respects he held the advantage. He was able to communicate officially and in private with his superiors who were all sympathetic to his cause. He had agents in the Regiment, and, as time went on, an increasing power to bring it home to others that they could either side with him or risk the fate of the Paymaster and the Sergeant-Majors. In short, he was now well placed to manoeuvre his opponent into a situation where a charge of peculation could be made against him.

His first move could well have been the mysterious robbery at the Pay Office. The disappearance of certain account books suggested that it had been staged by Smales himself to remove evidence of fraudulent activities he had been engaged in. This was the obvious explanation and the one most likely to be accepted by the authorities. But there was another which they would find it easy to ignore—that the break-in had been arranged by someone who wanted it to *appear* that the Paymaster was getting rid of incriminating evidence. Whatever the truth, the Mhow Authorities were at no great pains to apprehend the intruder and he was never found.

Whether Crawley had secret knowledge of the affair or not, he certainly made good use of it. For by putting the screws on Sergeant Bennett he had acquired a valuable agent in the enemy camp. Then on 27 May, with the approval of Sir William Mansfield, a committee with Colonel Prior as president and Renshaw and Garnett, two of Crawley's main supporters as members, was ordered to descend on the Pay Office to inspect all books and accounts.

It came up with the disappointing news that Smales was in credit by 45 Rs. But the next day all books and furniture, and according to Smales, some of his private property, were removed from the Pay Office and taken to the house of Riding Master Malone, who had been Acting Paymaster since the Court-Martial began. At this, Smales immediately disclaimed all responsibility for the Regimental Accounts since he could no longer exercise effective control over them.

On 2 June another move to isolate him came when a Regimental Order forbidding any N.C.O. or private to communicate with him was read out to the troops on parade.* Apart from the arbitrary nature of the Order, it was read out on three consecutive occasions which was unusual and naturally regarded by Smales as another attempt to degrade him before the Regiment while the Court-Martial was still in progress.

For a month following the Court's verdict, however, there came a pause in his harassment, the explanation for which seems to be that Crawley had suffered a breakdown of some sort. In an official letter dated 7 July he explained:

'It has been as I feared, and my bodily health has given way under the pressure which Paymaster Smales has so unceasingly brought to bear against me in the shape of false and malicious accusations heaped upon me, without the shadow of a foundation in truth for any one of them.'

Meantime Smales occupied himself feverishly writing long protests and appeals against his treatment during the Trial. Then, on 18 July he learned from Divisional Orders that the verdict had been approved by Sir Hugh Rose, the C.-in-C. India, and that he was cashiered. The news had been conveyed by telegraph in order to hasten the re-establishment of Crawley's authority, a method which was later to arouse comment by reason of its novelty. It left Smales a civilian entitled to no further pay or any of the advantages and protection enjoyed by a British Officer.

On the same day Malone gave up his appointment as Acting-Paymaster. He was replaced by Renshaw, and a second inspection of the accounts was ordered just as Smales had decided it was time to set off for home. What happened may be told in his own words:

'When I left Mhow on 21 July I did so from the fact of the finding and sentence on my Court-Martial having been published in Divisional Orders, Lieutenant-Colonel Crawley addressing me as 'Mr. Smales', intimating that no further issue of pay

---

* The high rate of illiteracy in the Army made this the only effective way of insuring that all soldiers were officially informed on all matters that concerned them.

6

would be granted me, and treating me, I think, unjustifiably, as de facto no longer connected with the 6th Dragoons.

'The whole proceeding, considering I had a large family in this country, living and travelling in India being of a most expensive character, and the distance from Mhow to the Presidency being so great, that it behoved me, if I contemplated having it in my power to remove myself and family at all, to do so at once. I did so, therefore, as soon as practicable after the proceedings in question had been put into force. I made no secret as to my movements. I had no cause for doing so. Lieutenant-Colonel Prior, Major Swindley, Surgeon Turnbull, Quartermaster Wooden, Assistant Surgeon McMaster, with other of my friends, were quite aware of my contemplated move.

'Everything I did, I did in the open day and in the most open manner, and I left Mhow by the Government mail-cart at 2 o'clock in the day. Notwithstanding these circumstances, the day after I left, Lieutenant-Colonel Crawley got up a hue and cry, and set a report afloat that I had "absconded, leaving heavy debts", and from which most unfounded and unjustifiable report I was arrested by the civil power on my road, but subsequently released upon the illegality of the proceedings being ascertained.'

The amount he was said to be in debt was the huge sum of £1,600. After his release he was allowed to continue his journey to Bombay where he boarded the S.S. *Malta* about to sail for England. But a few hours before the ship sailed he was again arrested and taken to gaol. (He does not say whether his wife and family also disembarked but the probabilities are they sailed without him.)

It was October before he was finally brought to court. He was prosecuted by the Government solicitor (who also happened to be Crawley's legal adviser). The Grand Jury, with an impartiality that contrasted sharply with the military justice he had just experienced, threw out the indictment against him.

The Authorities, in a last effort to delay his return home, insisted that references must be made to England before he could be allowed to leave the country. But that too failed. Denying any intention of trying to evade his legal obligations, he declared himself insolvent and at last obtained his freedom.

While he had been defending himself against the civil charge he had not lost sight of his main objective, that of getting the verdict passed on him at Mhow reversed. He sent off a steady stream of letters to London, some written in the 'common jail' into which he had been thrown, giving in great detail his version of all aspects of his Court-Martial and the subsequent events. In all he despatched seven to the Adjutant-General at the Horse Guards and three to the War Office.

He posted his last from India on 10 October. By 6 December he was in London posting two long 'Memorials' from his address, 126 Leadenhall Street.

No decision ever seems to have been reached as to who was responsible for the alleged deficit in the Inniskilling's Regimental Accounts, and the possibility has to be accepted that Smales may have feathered his own nest when he knew he had lost his case. But if he was a defaulter he certainly did not behave like one. It must also be said that the known facts tend to support the view that the affair was deliberately engineered to discredit him and so frustrate his intention of pleading his case before the authorities in England.

The Court of Inquiry that looked into his accounts at the end of May had found them in order and its two members, Renshaw and Garnett, were unlikely to have missed anything suspicious.

The balance sheet which later showed Smales to be in debt for £1,600 was drawn up by Bennett and Malone, both of whom were now so firmly under Crawley's control as to be mere puppets, the first for reasons already explained, the second because he knew nothing of accounts and would have been liable to charges of embezzlement himself unless he 'played ball'.

Malone was examined by the barrister defending Smales at Bombay in 12 August and made this statement:

'The Account "Q" (on which Smales was being indicted), has not been audited by anyone, but prepared by me and Sergeant Bennett. The account has not passed through the Divisional Pay Office at Mhow; it has not passed through any of the offices of military finance in Bombay; it has not even been passed by a regimental committee of the 6th Dragoons.'

He also admitted that he did not know very much of accounts

and added, 'There are many mysteries of a Regimental Pay Office which I have not yet fathomed. I do not understand Regimental Pay-Office Book-keeping.'

Prior and Swindley were also questioned by Smales' counsel. Their statements generally agreed, but as Prior seems to have managed to remain as impartial as anyone who became involved in the affairs of Crawley, his remarks may be given some credence.

He thought it had been optional whether the defendant should have attended the Committee that was investigating the accounts before he left Mhow. The Committee had not looked into the back accounts nor had it verified the deficiency of 16,000 Rs which, as far as he knew, depended upon the calculations of Mr Malone.

Any notice for Smales to attend would have gone through him as President and none had been sent. He did not think Smales had absconded as he had heard on good authority that he had left Mhow in the public mail-cart.

Smales had given every assistance to the Committee that made up the accounts to 27 May and he was satisfied with the way the accounts had been kept, as were the other members of the Committee.

Prior concluded: 'The Committee was held whilst the defendant was under arrest, and occupied in defending himself before a Court-Martial. I never heard of an officer being brought before a committee and at the same time defending himself before a Court-Martial.'

Unless Prior had been completely misled in the matter of the accounts his remarks point strongly to Smales' innocence. What he said must have been unpalatable to Crawley as well as to the Commanders at Mhow and Bombay. In particular, the part quoted was a direct criticism of Sir William Mansfield who ordered the first investigation of the Regimental Accounts when the Court-Martial was still sitting.

In view of what was happening to those who dared to cross their superiors at that moment, it must have required some moral courage to speak on behalf of one who had just been cashiered and might soon be subjected to further degradation.

12

The Top Brass confirm and comment

Before its final ratification, the verdict had to pass before two high ranking officers in India, Sir William Mansfield and Sir Hugh Rose. Their decisions and all the documents connected with the matter would then come under the notice of the Duke of Cambridge, the C.-in-C. the British Army, at the Horse Guards. The views of all three were to become of public interest and controversy.

In his handling of the Crawley business, Sir William Mansfield had probably acted no differently from most other military commanders of his day in manipulating the machinery of justice to maintain the authority of a commanding officer. But now that events had combined to expose him it was natural he should resort to casuistry to justify his actions.

On 5 July he forwarded the Proceedings of the Court-Martial, together with his comments, to Sir Hugh Rose for his approval, devoting a good deal of space to explaining those decisions not strictly in accordance with Army Regulations.

He had thought it advisable not to order the usual Court of Inquiry before Smales was tried because experience of what took place the previous November, when Smales was arrested by Crawley, showed that no inquiry would be of use unless matters were brought to a direct issue between the Paymaster and his C.O. on oath. The prolonged and bitter discussions it would inevitably have produced would have had the most injurious effects on the discipline of the Regiment. With or without an inquiry, a court-martial was inevitable, for Smales had 'forced a deadly quarrel' on Crawley, who had either to accept the challenge or place himself in Smales' hands and 'must ultimately have been himself compelled to quit Her Majesty's Service ignominiously'.

On the arrest of the Sergeant-Majors, Sir William wrote that they were animated by ill-feeling and a sense of grievance towards their C.O. The death of Lilley was much to be regretted. But even if this was not the direct result of strong drink, that must have accelerated it. For Lilley had not been allowed to communicate with others during his arrest and must have consumed the large quantities supplied to him.

Although Smales had attributed the worst motives to his Commanding Officer for arresting the Sergeant-Majors, once Colonel Crawley had 'established in his mind' that a conspiracy was organized against him by the N.C.O.s and that is was a 'mutinous attempt', he was literally without choice in the matter.

Having admitted that the dangers from the Sergeant-Majors existed only in Crawley's imagination, Sir William went on to add that, though he did not consider there were legal grounds to attribute more than gross impropriety to them, he thought it best to wait what would come out of cross-examination at the Trial before condoning the offence. Regarding the severity of the arrest, this was entirely due to the culpable neglect of the Adjutant.

He dismissed the taking of statements by Crawley from others present at conversations between Smales and himself as a simple precaution sometimes taken by commanding officers for their self-protection, but seems to have forgotten that on a previous occasion he had stated he was convinced the practice would be absolutely repugnant to Crawley. He added, however, that the statements were not taken *secretly* because Smales had found out about them.

As to the accusations that Crawley had intimidated witnesses during the Trial, the Court had invited Smales to produce anyone who had been. From the fact that he failed to, it could be assumed there were no grounds at all for the allegation. Having thus ignored that an intimidated witness was hardly likely to come into Court to say so, Sir William went on to assert that not only had Smales raked up the matter of intimidation after his

Defence was closed but had himself been 'practising espionage in the regiment'. This was why orders had been sent that he was not to be allowed to communicate with N.C.O.s or Privates until the verdict of the Court was known.

Crawley, who was at present too ill to attend to business, had admittedly at times been wanting in tact and judgement. But he had faced great difficulties by the opposition to his authority and the lack of support from his officers. The Defence had failed to produce any proof of foul language or oaths used by him to individuals and the instances of harsh conduct were, with a few exceptions, merely what took place every day on parade in enforcing attention and discipline.

Sir William said he would refrain from going into details about the conduct of some of the officers who gave evidence but it seemed they should be 'summarily removed from the Regiment'.

In a postscript to his remarks, the General showed he was not altogether happy about his part in the affair or the interpretation that might be placed on it in England. He believed the Duke of Cambridge wanted the fullest information on matters connected with the Court-Martial and he would deem it a great favour if Sir Hugh Rose would forward his report to H.R.H.

Six weeks later, Sir Hugh set his seal on all that had happened by issuing a very long and detailed General Order that ranged well outside the Court-Martial. And so at odds are his partisan and ill-conceived comments that only one rational explanation of his conduct seems possible—he was misled by the memory of his painful associations with the Inniskilling's officers into using the Court-Martial to vindicate his disapproval of their behaviour of two years before.

The proceedings furnished *proof* 'that before Colonel Crawley's arrival there was discord between officers of the regiment, neglect on the part of many of them, the two seniors included, of the simplest yet most essential cavalry duties, impatience of reproof, however merited, unaccompanied by amendment, which, under the baneful example, extended to the non-commissioned officers, and caballing of the one and the other against the head of the regiment'.

The criticisms made.by Colonel Crawley and the Regimental

Orders he had found it necessary to issue, were *proof* that the officers were neglecting their essential duties. His remarks on the riding of the Regiment displeased some of the officers but Sir Hugh had himself found fault with the equitation two years before and his opinion had been confirmed by a distinguished cavalry officer who inspected it just before Colonel Crawley arrived.*

It was however in his blind endorsement of the attacks by Crawley in his closing speech on the dead R.S.M. and individual officers, that Sir Hugh displayed the greatest irresponsibility and disregard for ordinary justice.

Of Swindley he wrote that on Crawley's arrival at Ahmednuggur, he 'literally refused his commanding officer, as president of the mess, the use of a few articles of mess plate which was necessary for him to eat his breakfast'.

And of Turnbull—he had been collecting evidence in support of Smales even before the charges had been framed. It was also hard to understand how, in stating the causes of Lilley's death, all of which tended to inculpate Colonel Crawley, he had failed to be aware of the large amounts of intoxicating drink bought by the R.S.M. while he was under arrest, for they were quite enough, in combination with Indian Summer heat, to compromise the life of a man in exercise and rude health.

As to Lieutenant FitzSimon, the annoyance caused by the posting of a sentry over the R.S.M. was entirely due to his carelessness and neglect.

Colonel Crawley had tried hard to make the Regiment efficient and the very favourable half-yearly Inspection by Major-General Farrell showed that he was 'an officer of much intelligence, energetic zeal and fully qualified for the command of a cavalry regiment'. Far from being too severe, he had on many occasions been too lenient when his authority had been questioned or resisted.

It was a great satisfaction to Sir Hugh to know that several officers had given their Commanding Officer cordial support and

* The officer was of course Brigadier Hobson, who, Sir Hugh omitted to add, reinspected the Regiment a few days after Crawley arrived to find the horses in first-rate condition and the movements on parade 'executed with celerity'. Curiously, the *only* fault found on this occasion was the backwardness of the children attending the Regimental School, whose education was very much neglected.

that the men had maintained their discipline like good soldiers. He ended by saying that he would not fail to bring the conduct of the three senior officers—Major Swindley, Captain Weir and Surgeon Turnbull to the notice of H.R.H. the General Commander-in-Chief.

Before leaving these pronouncements it is worth recalling that when the Inniskillings had been commanded by Colonel Shute, Sir Hugh had blamed the disagreements among its officers on that C.O.'s lack of proper control over them. He seems unaware of any inconsistency in absolving from blame another C.O. who had himself become the centre of far worse dissentions involving nearly all the officers of the Regiment.

By the end of November, when ex-Paymaster Smales was making his way home to England with all speed, the Proceedings and all the documents relating to his Court-Martial had arrived and were being critically examined at the War Office and the Horse Guards.

The responsibility for delivering final judgement on the whole affair would rest on the C.-in-C., the Duke of Cambridge, who, being thousands of miles away from the scene of events, was able to take a more detached view than had the Authorities in India. Ideally he could bring in a verdict based on the strictest rules of justice; but in practice he would have to balance this with the fact that failure on his part to support the stand taken by those on the spot could have harmful effects on the discipline of the whole Army.

Nor was he entirely a free agent in other respects. He could be answerable to the War Office, to Parliament and in the last analysis, to the British public for his decisions. He received this advice from his chief legal adviser, Thomas Headlam, the Judge Advocate General (at this time a member of the Government in the Commons).

Had the Court-Martial on Paymaster Smales taken place in England, and had his opinion been asked, he (Headlam) would have advised that the charge against the prisoner should be confined solely to the writing of the letter to his Commanding Officer. The only evidence needed to support it would then have been proof of the prisoner's handwriting. If the accusations made in the letter were thought to have any foundation, they could have been made the subject of a separate investigation.

Smales had been charged with making false and malicious accusations, but not with knowing them to be false. It was therefore quite consistent with the charge that he believed them to be true, in which case he could hardly have been accused at the same time of behaviour 'most disgraceful and unbecoming the character of an officer and a gentleman'.

These remarks on the way the charges had been framed were no mere technicality, for the error had affected the whole Trial, making it uncertain what precise crime the prisoner was guilty of. It had resulted in the prosecutor and the prisoner being both accuser and accused, that is, two persons being tried at the same time 'with machinery only fitted for the trial of one'.

Had the trial taken place in England he would have advised a re-trial on the charge stated. But in the present situation that procedure could not be adopted. *The Proceeding of the Mhow Court-Martial, though objectionable, did not seem to be illegal, and the prisoner was undoubtedly guilty of a serious military crime in writing the letter.*

Armed with this advice, the Duke of Cambridge proceeded to apply it to the construction of a Memorandum which he probably hoped would be restricted to as few readers as possible.

He had seldom found himself in a position where it appeared more difficult to deal out justice to all parties while doing justice to the Service. Not only was a most distinguished Regiment implicated, but he was forced to disagree with the views of officers of high rank, great reputation and usually of excellent judgement.

He had, however, but one course to pursue, which was, after mature consideration and a patient hearing of the opinions of those most likely to give unbiased advice, to act upon his own sense of justice.

He was sure the Court had come to the proper verdict as regards the insubordinate tone of the letter written by Paymaster Smales, and his removal from the Army was an act of justice to the Service. Unfortunately, however, in bringing down on his own head a just retribution, his trial had exposed a state of affairs in the Inniskilling Dragoons which he would try to view impartially.

Under Colonels White and Shute—two distinguished officers in whom he had great confidence—the Regiment, before they embarked for India, were all that could be desired as to esprit de corps and good feeling amongst the officers, and as to drill and discipline among the men. On their embarkation some changes took place among the officers, and not long after, some unfavourable reports of the behaviour of one or two individuals when off duty called for his severe displeasure.

Still, that the Regiment remained in the highest state of discipline was shown by all the confidential reports that had reached him until Colonel Shute was succeeded by Colonel Crawley, an officer experienced in the lower ranks of the Service, of considerable talent and zeal, but unfortunately not gifted with the special talent which united the firmness of command with the tact which inspired confidence and good-will.*

From the first he appeared to have taken an unfavourable view of some points in the Regiment and to have expressed himself in no measured terms as to the changes he contemplated, which he (the Duke) could not but think was uncalled for and was sure to create an unfavourable feeling in the Regiment. His conduct after the Court-Martial, if the address he made to the Regiment had been correctly reported, was, to say the least, irresponsible and injudicious.[7]

He could not speak too strongly of the confinement, under arrest, of certain non-commissioned officers during the Trial on a charge of conspiracy which was never attempted to be proved against them. He believed that if the C.-in-C. in India had been better acquainted with some of the facts of Sergeant-Major Lilley's case, he would not have attributed his death to excessive drinking.

Only the high opinion expressed of Colonel Crawley by the General Officers would induce him to continue that officer at the head of the Regiment. He did so only upon trial and in the hope that he would carry on discipline without outraging the feelings of the gentlemen under his command.

The tone and manner in which Major Swindley had given his evidence was highly unbecoming and were it not for the remarks passed on Colonel Crawley he would at once be removed from

* An apt assessment.

the Regiment. He too would be given the advantage of previous reports in his favour.

As to Captain Weir and Surgeon Turnbull, nothing could excuse a subordinate officer from showing his disapprobation of his commanding officer before younger members of the corps.

In conclusion he trusted that the future conduct of the officers of the Inniskillings would eradicate the evil spirit which appeared momentarily to have crept in to tarnish a reputation that was second to none.

13

Mr Smales widens the conflict

From the moment he arrived in England early in December, Smales pressed on vigorously with his campaign to get the verdict passed on him reversed. And whatever the rights and wrongs of his case, there can be no doubt as to the courage and determination with which he set about it, or indeed of his remarkable physical and mental stamina.

He was now 53 years of age and had spent more than half his life in disease-ridden lands and climates which normally exacted a crippling if not fatal toll on those who lived there for even short periods. Yet he seems to have emerged with health and mental faculties in no way impaired. Nor does he appear to have been affected by the emotional strains of his long battle of wills with Crawley, his extended trial, his imprisonment in a Bombay gaol not to mention his financial and family responsibilities.

In India he had fought for many months what was virtually the whole Army Establishment; now, undeterred by his defeat, he was preparing to take on the even more august top brass in England. He was not without friends, but the kind of friendship he needed was of a rare quality—how rare, his recent experiences had shown all too clearly.

For most of his allies, his cause was a lost one. Those who had already yielded to pressures from his opponent could hardly be expected to re-enter the arena and completely wreck their careers by openly coming out in his support again. Others would only do so if goaded by self-interest.

He does not seem to have won the kind of partisan warmth

that Crawley was able to rouse. He sometimes referred in his correspondence to 'my friends'. None of these individuals appears to have returned the compliment. His testimonials vouch for his intelligence and business acumen, but even those referees who knew him for long periods display no real warmth towards him.

Many individuals—Swindley, Turnbull, FitzSimon, Shute and others—all had good reasons to feel aggrieved after what had happened, but their appeals and protests would all conform to conventional behaviour. (The two Sergeant-Majors, Wakefield and Duval, of course had virtually no effective means of protesting at all.) But the approach of Smales had, from the start, been less circumscribed. His logical mind had always tended to break out of the accepted code of conduct. Now, with nothing left to lose, he had no inhibitions about stepping outside it altogether and insisting that the ordinary rules of civil law and justice must be applied to his case. He was ready to go it alone, but he could, as events were to show, become the focus and guiding force of the attacks of others.

It is unlikely that anyone at the Horse Guards saw in him anything like the pervading malevolence that Crawley described in his phrase 'upas-like shadow'. Yet they could hardly fail to suspect his hand in the assaults that gradually developed in different quarters, all aimed at tearing down the wall of secrecy behind which they hoped to hide the unsavoury events that had occurred at Mhow.

His official strategy is disclosed in the steady flow of correspondence with the Military Secretary to the Duke of Cambridge and the Adjutant General.* He presents his case in detail, probing at every joint in his opponents' armour of official reticence. He records in his diary every remark and admission likely to be of use. He never lets up.

His private correspondence, too, must have been extensive—to India and to interested parties at home. He presumably wrote also to persons of influence who might be persuaded to take up his cause. But perhaps because of the accusations against him of peculation, a long time was to elapse before anyone openly declared his support.

* General Yorke Scarlett, who led the successful charge of the Heavy Brigade at Balaclava.

The lines of attack that were to prove most effective are set down in a letter dated 6 December, 1862, and developed in another on 4 March the following year. With the second he enclosed letters and documents he had received from officers involved with him in the dispute with Crawley and who were naturally debarred from communicating directly with the Horse Guards or War Office. To this extent his civilian status placed him at a definite advantage both in fighting his own case and as a forwarding agent for others.

By early March his case stood as follows.

He was pressing for an investigation into the circumstances in which Lilley had died, on the grounds that the R.S.M. had been an important witness for the Defence and also because he felt a personal responsibility as the innocent cause of the man's death. To support his claim that it had resulted from the nature of the arrest and not as Crawley asserted from an excess of drink, he produced a good deal of written evidence in Medical Reports and letters from Turnbull and his assistant, Dr Barnett. These all pointed to the conclusion that Lilley could have consumed little of the drink supplied to his house, since most of it had been medically prescribed for his sick wife.

He also forwarded a copy of an amended bill showing the amounts of intoxicating drink and other items delivered to Lilley's house. This had been sent to him by Turnbull and bore the signature of the same Indian merchant who had allegedly given the first to Crawley. It listed only twelve bottles of brandy instead of 23 as in the original. With the bill was a signed statement by the merchant to the effect that Mrs Lilley had settled this account after her husband's death and that he knew for certain, that he had been the sole supplier of drink to the house during the R.S.M.'s arrest.

Regarding these bills, Smales wrote: 'Letters are now in England and can be produced, proving that the account of the liquor upon which Sir Hugh Rose was misled, was obtained by means the most discreditable, and was to all intents a forgery; it was not prepared by the merchant who supplied the liquor, but his book was seized by Lieutenant Davies and carried off to Lieutenant-Colonel Crawley's house, where the account in question was prepared.'

Another enclosure of peculiar interest from being of a private nature, was part of a letter apparently written by Turnbull to Colonel FitzWygram, now commanding the 15th Hussars in England. How Smales came in possession of it, he does not say. It was dated Mhow, 23 January, 1863, and read:

'With regard to Sergeant-Major Lilley's case, Colonel Crawley has done all he could to prove he was a drunkard and was furious with me because I would not side with his views. Dr Barnett and I frequently saw him during the period of his arrest, and never saw him the least the worse for drink. I believe the poor fellow died of a broken heart from the ill-usage he received from the Mhow and regimental authorities. I wish I could have placed this on paper and on official record, but the arbitrary government we live under out here prevents me doing so.

'In my addenda to Sergeant-Major Lilley's case I fear I have not expressed myself clearly. I did not mean to say that Dr Barnett and I had found Colonel Crawley's statement about the consumption of liquor to a large extent by Lilley to be correct, but that Mrs Lilley said he was in the habit of taking a fair quantity daily, but she could give no definite idea of the quantity. I believe the Acting Adjutant (Davies) must have frightened the poor woman to make some admission, or perhaps it was merely trumped up by Davies and the Parsee that Lilley had got the quantity stated.

'Every one of the sentries who were placed over him day and night, and who were examined by Colonel Crawley, stated that they had never seen him in the least intoxicated, or seen brandy taken into the room. Colonel Crawley stated to me in the orderly-room that he was aware from the day he joined that Lilley was a sot. I remarked that if he thought so that it was strange that he should have entrusted him with the charge of the wines &c., in both the officers' and the sergeants' mess, that I had always considered him a most sober man, and respected by everyone in the regiment.

'We who have spoken the truth feel as if we stood over a volcano, &c.'

Also bearing upon the circumstances of Lilley's close arrest were two letters written by FitzSimon denying the statement by Crawley that the manner in which the sentries had been placed in the R.S.M.'s house was due to his carelessness and neglect of duty. The first, addressed apparently to Smales, contains the following:

'I can easily prove, although Colonel Crawley will not admit it, that he was perfectly aware that Mrs. Lilley must have been inconvenienced by the sentry, for, at the time he gave me the order to place the sentries so that they would not lose sight of the prisoner night or day, Sergeant-Major Cotton remarked to him that poor Lilley was a married man; and to the best of my recollection, I said at the time, that Mrs. Lilley was sick, and her husband was obliged to rub some ointment on her chest every day; it was then he flew into such a rage, and said he did not care a d—— whether he was married or single, he would have the duty done.

'Sergeant-Major Lilley while under arrest was changed from his staff quarters into that of a Troop Sergeant-Major's, which was composed of only one room; and as the Colonel was aware that Mrs. Lilley was with her husband, I do not see how he can say that he did not know that she was inconvenienced by the sentry. I saw and visited Lilley every day, generally twice, and can swear that I never saw him under the influence of liquor; I always found him to be most abstemious.'

In the second letter, dated 3 November, to Davies, the Acting Adjutant, FitzSimon began by requesting that its contents should be laid before Sir Hugh Rose. He went on to deny responsibility for the way in which the sentries had been posted and to refute the statement by Crawley that he had poor eyesight.

'I have, lastly, respectfully to observe, that Lieutenant-Colonel Crawley, in his reply to the Court-Martial, observes that I was so blind that I could not distinguish any person at five yards distance so as to be sure of his identity. I most respectfully beg to assure his Excellency Sir Hugh Rose that I am not blind, and three Medical Officers who have examined my eyes, testify that there is no abnormal appearance which might lead any person to suppose that I was blind.'

In support of this FitzSimon produced medical certificates from Turnbull and McMaster, two M.O.s of the Regiment, (Barnett was on leave). Both certified they had tested his long and short sight and found them 'very good'. Smales, with his careful attention to detail, had also written to Colonel Shute, now commanding the 4th Dragoon Guards, and received a reply from him to the effect that he was sure FitzSimon had no defect in his eyesight when serving under him as Adjutant in the 6th Dragoons. But what must finally to have demolished Crawley's case was the fact that FitzSimon had gone on the rifle range and qualified as a First Class Shot at 500 and 600 yards, the only officer in the Regiment to hold this distinction.

Although the case put up by Smales thus far was a strong one, it could hardly have persuaded the home authorities to alter the verdict of his Court-Martial. However much they might have deplored the illegalities and injustices that had obviously taken place at Mhow, prudence supported by precedent demanded that the Commanding Officer who had perpetrated them and even more so the superiors who had supported him, must be shielded from public censure.

But apart from the persistent Smales, they had already been given some cause for uneasiness. The distasteful prospect of having to take some positive action had appeared momentarily on the last day of December, when both the Secretary of State for War and the Duke of Cambridge received a 'Humble Memorial' (from some god-forsaken township named Spilsby in Lincolnshire), bearing the signature of Richard Lilley aged 76 and the 'mark' of his wife, Mildred, aged 73.

The two signatories, the aged parents of the late R.S.M., after claiming to have been mainly dependent for their subsistence on the financial help given them by their dead son, continued:

'Your Memorialists humbly state their belief that it can be readily and most easily proved, that by the wicked and illegal imprisonment of their son he was deprived of his life, and that the treatment he received at the hands of his Commanding Officer, which caused his death, was directly contrary to the laws of his country, and of the Army.

'Your Memorialists humble crave your Excellency graciously to recommend that your necessitious Petitioners, in consideration of their very distressing position, may be permitted, for the few remaining years they may have to live, to receive the pension their late beloved son would have been entitled to have received, had his life been spared for only two years longer, which your Memorialists sincerely believe, under Providence,

would have been the case, had he not been subjected to his illegal imprisonment. . . .'

The petition was considered by the War Office for three days,
then turned down on the grounds that no funds were available
for paying it, a decision perhaps not entirely due to stony hearts.
The granting of a pension in such circumstances would certainly
have been seen by the vigilant Smales as an admission of responsibility by the Army for Lilley's death. Or perhaps the War
Office suspected the ex-Paymaster of initiating the appeal for
that very purpose, and indeed, what followed a month later lends
support to this supposition. For on 2 February the Military
Secretary received the first of a series of letters from the solici-
tors acting for the Lilleys. The firm, Thorndike and Smith, was
later revealed as the legal representatives of Smales and could
well have been advising him since his arrival in England.

At this stage, however, they demanded access to all documents
relating to the death of Lilley. They were met by the usual delay-
ing tactics; the War Office and the Horse Guards passing the
correspondence from one to the other, and sending no reply
before reminders were received. Finally the firm was told that
the papers had not arrived—a palpable falsehood, since they had
certainly been in the hands of the Duke of Cambridge when he
composed his remarks on the Court-Martial.

After four weeks badgering, Turnbull's Medical Report suddenly turned up at the War Office and was sent to Thorndike and
Smith. They promptly requested further information including
an official letter from the Surgeon in which he re-stated his
belief that Lilley's death had not been hastened by drink.

After more delay they were given part of the letter. They
immediately wrote demanding the recall of Crawley, saying they
intended to have him charged with manslaughter as soon as he
set foot in England.

Obviously alarmed at the turn of events, the War Office began
debating with the Horse Guards the advisability of having
Lilley's death investigated on the spot in India, no doubt hoping
that some more plausible explanations than those already received would be forthcoming. But while they were deliberating
a fresh jolt to their complacency appeared suddenly from yet another quarter.

About the middle of March, a pamphlet, written by a Colonel Hugh Walmsley, the son of a well-known Radical MP from Liverpool, and addressed to the members of the House of Commons, began to circulate. A version which appeared in the *Liverpool Daily Post* of 14 March, 1863, went:

Military Despotism: or the Inniskilling Dragoons.
A Tale of Indian Life.

'An extraordinary tale has just been published in the form of a pamphlet, the contents of which appear to be fully authenticated. It has been written by a Colonel at the request of one who tells his story as follows:

'I am a poor man, a working painter by trade, living at Llandudno. My parents are old people residing in Lincolnshire. I had a brother, now dead, who served for nearly 19 years in the Inniskilling Dragoons. He was a kind, good man, an honour to his family and the chief support of our aged father and mother.

'I subjoin to this pamphlet a certificate* given by his late commanding officer, speaking of him in terms that I, as his brother, feel proud and grateful for. After 19 years good and faithful service to his country, my brother suddenly found himself brought up before his then commanding officer, not in the orderly room, but in Colonel Crawley's private house. A vague charge was brought against him, the truth of which he to the day of his death strenuously denied. There were no witnesses to prove this charge, but, nevertheless, my brother who had been summoned as a witness on the defence of Captain Smales, against whom Colonel Crawley had preferred charges, was marched to his quarters, a close prisoner.

'Thus he was detained in his own room, a sentry duly armed keeping guard over him. His wife was dying at the time, but

* 'I have very great pleasure in testifying the very high opinion I have ever entertained of Sergeant-Major Lilley. I knew him well during his whole service in the Army. He was for a long time in my troop when I was Captain, and was R.S.M. the whole time I commanded the regiment. I consider him one of the most straightforward, truthful and worthy men I ever knew; thoroughly sober and trustworthy, an excellent soldier, and respected by all who knew him.
            C. Shute, Colonel 4th Dragoon Guards.
Shute had been selected by the Horse Guards to reorganize the 4th Dragoon Guards after the Bentinck–Robertson scandals.[8]

nevertheless the armed sentry was placed in the dying woman's room. No charge was ever brought against my brother. He died after weeks of close confinement under an Indian sun. His wife died also—his children died—his two comrades who were also made prisoners at the time, and on the same unsubstantial and vague charge, lost one his reason, the other his bodily strength. My brother then being dead, they were released after ten days further confinement, without trial, and without any satisfaction being given them.

Samuel Lilley

This introduction was followed by the first of many accounts, varying only in detail and literary style, that were soon to flood the newspapers and place the events of the Mhow Court-Martial before the entire reading public of Britain.

The rising concern of the Military Authorities at the way events were shaping is shown in their communication to Thorndike and Smith of 21 March informing them that orders had been sent to India for the circumstances of Lilley's death to be further investigated.

Then in response to the plea for justice by the dead man's brother, on 23 March, Captain Archdall, MP, an ex-officer of the Inniskillings, asked the Secretary for War (Sir George Lewis), whether the attention of the War Department or the Horse Guards had been drawn to the recent allegations in the press and elsewhere on the confinement of R.S.M. Lilley during the recent Court-Martial at Mhow. Had the inquiry taken place and if so, with what result?

He also wanted to know whether any more information was available regarding the remarks made by Sir Hugh Rose on the state of the Inniskilling Dragoons and the death of R.S.M. Lilley being caused by 'ardent spirits' rather than his confinement.

The Secretary replied that the report from India had been called for and he believed it was now at the Horse Guards. The matter was still under consideration.

In this he was either himself confused or was misleading the House. It is true that a report by Sir William Mansfield on the arrest of the Sergeant-Majors had just been received but this was not of course the investigation that had just been ordered. No doubt the Secretary hoped to give the impression that the War Office had matters well under control.

This was certainly not the case, for the Military Authorities continued their tactics of delay and passing the buck perhaps hoping the rumours now sweeping the country would die down.

At the same time they were made to realize that now Smales had found a breach in their defences he would not neglect exploiting it to advantage. In an attempt to force their hand he wrote challenging the right of Sir Hugh Rose to confirm the verdict against him, contending that the Queen alone had that power. The Horse Guards replied with a show of firmness, telling him that the C.-in-C. in India could confirm verdicts of courts-martial, that he was cashiered and that another paymaster was being appointed in his place. Smales protested. But events were now moving so rapidly that his attempts to influence his fate were becoming almost irrelevant.

14

J.O. intervenes

On 11 May a new champion announced his entry into the fray with a letter to *The Times*. It was signed 'A. Civilian', but this, far from hiding the writer's identity, revealed it at once to all the paper's regular readers as one James Matthew Higgins, popularly known as 'J.O.'.[9]

The following extracts from this, his first letter on the Crawley affair gives some indication of the style and powers of advocacy that made him the most formidable controversialist of the day on burning social issues.

He begins by referring to a Royal Commission which had just been set up to enquire into grievances by certain 'distinguished service Colonels in order to dispel the idea that the Government would treat any class of military men with neglect or injustice.'

He goes on:

'It is impossible not to applaud the over-scrupulous motives by which Lord Palmerston has been actuated in the matter. The duties of the members of the Royal Commission thus appointed to inquire into and decide upon what has already been fully inquired into and fairly decided upon will, necessarily, be very light; and as they are all able and good men it seems a pity to allow them to separate before a little serious and useful work has been got out of them.

'Now there is another military grievance which has acquired a wide notoriety, and which has done, in my humble opinion, even more than the case of the "distinguished service Colonels" is likely to do towards instilling into the public mind the idea

103

that gross injustice is often done in the military service to very deserving individuals with impunity; and if the offender happens to belong to the upper classes and the victim to the lower classes, it is almost impossible to obtain for the latter any redress. I would therefore suggest that the Royal Commission, appointed to inquire into a matter which has already been inquired into, should be directed to inquire into the circumstances that led to the death of John Lilley, Regimental Sergeant-Major of the 6th Dragoons at Mhow, on 25 May, 1862.

'If the allegation made on behalf of the "distinguished service Colonels" are correct, these unfortunate officers, after having been liberally rewarded for the services they had rendered, have since been deprived of a very small portion of the reward they received; their alleged loss being estimated by themselves at £400 each. If the allegations made on behalf of Sergeant-Major Lilley be accurate, a gallant and good soldier has been done to death by those set in authority over him. Surely the one case calls for investigation by a Royal Commission as the other.'

(Then follows a brief account of Lilley's military career, the circumstances that led to his arrest 'to prevent his having communication with anybody else', his sufferings in confinement leading to his death and the attempt by Crawley to impute this to drunkenness.)

The letter concludes:

'It is impossible to read this document bearing on John Lilley's deplorable case without seeing that in the course of the prosecution to which the poor fellow was subjected, every rule of the service established for the protection of the inferior against the superior was infringed by Colonel Crawley, who, nevertheless, is still in command of the regiment in which this tragedy was performed: I will not however enter into these details; I had rather leave them to be noticed by the Royal Commission about to meet. I feel sure that her Majesty's ministers will be unanimously of the opinion that such a case as this is at least as worthy of investigation as are the petty grievances of 21 "distinguished service colonels", which, having been investigated, are about to be investigated again, "lest the public should imagine that an injustice is ever committed with impunity towards any class of military men deserving of consideration".'

Within twenty-four hours of the publication of J.O.'s attack, the War Office became galvanized into taking a step which it could have taken at any time in the past six months—it asked the Judge Advocate General for his opinion on whether the arrest of the Sergeant-Majors had been legal.

That official had, it may be recalled, decided the previous November that the Court-Martial on Smales, while objectionable, was not illegal. He had admitted the numerous and glaring irregularities that had characterized the whole proceedings, but nevertheless decided the Court had reached the correct verdict since Smales was obviously guilty of insubordination in writing the letter.

He must have known of course that this decision would not stand up to impartial scrutiny for the simple reason that Smales had not been charged with that offence but with making statements in his letter that were false and malicious. But at that time he was probably not over-concerned with the niceties of Court-Martial procedure or with military justice but with assisting the Authorities out of a delicate situation and preventing a military scandal.

Now, however, with public resentment over the death of Lilley likely at any moment to erupt into Parliament, the question of the arrest of the Sergeant-Majors had become of major importance. His decision on that issue had not only to be legally sound but must be seen to be so. He took nine days to reach it and on 21 May he despatched it to the War Office.

He began by summarizing all the material evidence on the arrest contained in the papers he had read and concluded that Sir William Mansfield was right in deciding that Colonel Crawley had failed to show sufficient grounds for bringing the Sergeant-Majors to trial for conspiracy.

From that it followed that they ought not to have been arrested in the first place. They could therefore have complained that their arrest was 'irregular and unauthorized'. Even if viewed in its most favourable light it was difficult to regard it otherwise than as an error of judgement by Colonel Crawley and one that was shared by General Farrell.

But Sir William Mansfield's decisions were even less explicable. He seemed to have seen nothing wrong in the arrests and even to have regarded them as 'justifiable and right'. He had decided there were no grounds for bringing the N.C.O.s to trial, yet gave orders that their arrests should be continued.

The reasons he gave were these:

1. To prevent the N.C.O.s being exposed to the temptation of conspiracy against their Commanding Officer.

2. To prevent their being tampered with.

3. That in the interests of justice the charge of conspiracy should not be condoned by their release until the trial of Paymaster Smales was over.

As Sir William had already decided there were inadequate grounds for the original arrest, its continuance must therefore be treated as a fresh arrest on new and distinct grounds. The question to be answered was whether these new grounds were legal.

1. The first reason given was obviously not. For there was nothing in Military Law that allowed a Commanding Officer to imprison a man to prevent him from being tempted to commit a crime.

2. It was difficult to understand what Sir William meant by 'being tampered with'. But if he intended they were to be kept in close arrest to prevent Paymaster Smales from communicating with them as witnesses in his defence, then this was 'not only illegal but an act of arbitrary power having for its object an interference with the administration of justice in a trial then pending, in favour of one of the parties concerned in that trial'.

3. If Sir William considered the three N.C.O.s ought to be detained while the trial of Smales was going on, in case some further evidence might emerge to support the hitherto untenable charge of conspiracy, then that too was no justification for continuing the arrest. Sir William may possibly have supposed he was acting as a magistrate remanding a prisoner in order to give the prosecution time to produce a known piece of evidence against him. But this analogy was so obviously incorrect as to be hardly worth considering.

From all this it was clear that none of the reasons given by Sir William Mansfield for continuing the arrest was strong enough to support its legality. The arrest could not therefore be justified 'under the Mutiny Act or the Articles of War, or the usages of Military Service'.

This advice from their legal expert obviously opened up for the War Office several lines of action against those who were responsible for the illegal arrest of the Sergeant-Majors should it become necessary to placate their increasingly vocal critics. But however they might decide to apportion the blame, they could be in no doubt that by sanctioning the arrest Sir William Mansfield had assumed responsibility for the actions of his subordinates and was therefore the most culpable of all the high-ranking officers concerned. Whether the War Office would be forced to recognize this publicly and take the appropriate steps against the General remained to be seen.

On the public scene nothing of importance occurred for the next few days but letters from the indefatigable Smales continued to arrive at one or other of the Military Establishments almost daily. In one he wrote that S.M. Wakefield was now in England, and Lieutenant FitzSimon had just arrived 'per overland mail'. Could they be officially examined on matters concerning them?

And in another; did H.R.H. still refuse to have his case laid before the Judge Advocate General?

Indicative of the pressures being exerted, the Military Secretary, having only a few days before refused the same request, now conceded that the Court-Martial proceedings had already been sent to that official for his appraisal.

Then on 29 May J.O. delivered a second broadside through the correspondence columns of *The Times*. He had received the proceedings of the Mhow Court-Martial and now presented the case against Crawley, explaining how the arrest of the Sergeant-Majors had been managed in order to discredit the evidence they were to give in the defence of Smales.

The only evidence produced by Crawley to prove that Lilley's death was due to drink was the bill sent in by a native dealer some time after the event. 'There was no proof that they (the spirits) had been sent, beyond the native's claim; there was no proof that Lilley had consumed any portion of them himself; no proof indeed that they had been consumed at all.'

Lilley had had an unblemished record for sobriety throughout his 18 years service and no one who was with him during his long arrest saw any signs of inebriety. When Crawley had him under cross-examination he asked no questions to suggest there was any doubt as to his sobriety.

It appeared later that Lilley's wife, in the last stages of consumption, had been kept alive for many days by large doses of wine and spirits, and the quantities prescribed for her by a Dr Barnett tallied as nearly as possible with that for which payment had been claimed by the Parsee wine and spirit dealer. Nevertheless, Crawley had sent information which misled the Commander-in-Chief into branding as a suicidal maniac one of the most deserving soldiers in the Service.

In his closing speech, Crawley had professed to being shocked when he learnt how Mrs Lilley had been treated in the posting of the sentries and went on to blame the Adjutant for the way it had been carried out, accusing him of having misunderstood his orders. But when Crawley had his Adjutant under examination he carefully avoided any allusion to the subject.

On 3 November Lieutenant FitzSimon, who had been severely censured in the General Order by Sir Hugh Rose, attempted to defend himself by addressing a letter to the General explaining that he had received explicit orders from Colonel Crawley as to how the sentries were to be posted and said he could name witnesses to prove it. But General Farrell, before whom the letter had to pass declared it to be insubordinate and directed it to be withdrawn. FitzSimon then wrote another letter saying he had been instructed by Farrell to withdraw the first; but this too he he was instructed to withdraw. The circumstances of this whole miserable transaction must be inquired into, said J.O.

All this and far more had happened, not at Naples under the Sardinians, but to one of the worthiest and best soldiers in the British Army, in British India.

John Lilley and his wife and both his children were dead. His parents were poor and obscure people; and if the outline he had given of this terrible case did not raise up for them assistance in some powerful quarter, they would themselves be quite powerless in bringing to justice those who had committee this foul deed.

On the day that J.O.'s second letter appeared, Mr Dudley Fortescue, MP for Andover, gave notice in the Commons that he would bring the whole circumstances connected with the death of

R.S.M. Lilley before the House that day week. Captain Archdall also asked for the Memorandum on the Mhow Court-Martial by the Duke of Cambridge to be laid on the table of the House.

At the Horse Guards and the War Office the large amount of official correspondence on the subject flowing in many directions shows the apprehension felt in both these establishments. Of this, two of the most significant may be mentioned.

The voluminous letters received from Smales on his Trial over the past months were forwarded to the Judge Advocate General for his views on the Court-Martial in the light of his decision that the Sergeant-Majors had been illegally arrested. And the now six-month-old appeal by Lilley's aged parents was sent to the Treasury with the suggestion that they should be granted a pension.

15

Parliament debates

Dudley Fortescue,[10] the Liberal Member for Andover, initiating the Parliamentary debate on the death of Lilley, reviewed the main events surrounding the Court-Martial. The source of some of his statements are somewhat obscure and some of the language he used to express them no doubt indicates the depth to which he was moved by the tragedy of Lilley's death. This with his known integrity and disinterestedness were to have a profound effect on the opinions of others.

He stressed that only on rare occasions when the charges were very grave or criminal was close arrest applied to Officers or Sergeant-Majors. Yet these three were subjected to the latter on charges of conspiracy, of which the Memorandum of H.R.H. the Commander-in-Chief stated there was not a shadow of evidence brought forward.

The 99th Article of War states:

'When a soldier or officer commits an offence deserving of punishment, he shall be ordered into arrest or confinement unless released by lawful authority, and no officer or soldier shall continue in such arrest or confinement more than eight days or until such time as a court-martial can conveniently be held.

And the 79th Article states:

'Whoever shall unnecessarily detain a prisoner in confinement without bringing him to trial, shall, if an officer, be liable to be cashiered, or suffer such punishment as by the judgement of a general court-martial be awarded.

(Loud cries of 'Hear, Hear'.)

But these men were kept in confinement, not for eight days, but for several weeks; one was liberated by death, and when the order came for release of the other two, it found one of them a raving lunatic, and he was conveyed to the hospital suffering from brain fever. After four weeks' imprisonment Sergeant-Major Lilley was taken ill and he died within a few hours. No inquest or committee of inquiry, which would there be the equivalent of an inquest was held, but a post mortem was made by the regimental surgeon who certified death due to apoplexy, brought on by his sedentary life and the peculiar circumstances in which he was placed.

The place where Sergeant-Major Lilley was confined was a single room in a bomb-proof building, formerly used as a cavalry stables, and which had since been pulled down as unfit for the occupation of troops. It was impossible that the roof of the building could get cool and the amount of heat collected during the day could not be carried off by night. In a room then, more like an oven than a human habitation was Sergeant-Major Lilley imprisoned, and this room was shared by his wife who was confined to her bed by diarrhoea attending the last stages of consumption. She only survived her husband by two or three weeks.

A sentinal was placed over the Sergeant-Major with injunctions to keep him constantly in sight and not allow him to receive any communication from without. In consequence of a sergeant's wife having attempted to communicate with the sick woman, Colonel Crawley, incredible as it might appear, ordered the sentry to be stationed inside the room, and there, in the presence of strange men, renewed from day to day, and posted three feet from the bed, all the functions of nature had to be performed by this dying woman. ('Hear, Hear'.)

Colonel Crawley pleaded ignorance of these circumstances, professed to be shocked when he heard of them and attempted to throw the blame of the indecent and inhuman proceedings upon the adjutant who received the orders. But there were witnesses to prove that it was distinctly brought to his notice that Lilley was a married man and that his wife was ill, and that the Colonel cruelly answered he did not care—married or single, officer or soldier, the duty should be done. ('Hear, Hear'.)

These facts were not imaginary but could be proved by witnesses now within the House. (Fortescue later commented that he had been mis-reported and that what he actually said was 'within the Country'.)

He then explained how slender was the evidence that Lilley had hastened his death by drink and said that His Royal Highness had endorsed the dead man's sobriety and vindicated his memory.

For redress and reparation it was unfortunately too late. Lilley was dead, his wife was dead, his children were dead. But if it was too late for reparation, it was not too late for punishment. (Loud Cheers.) Such punishment as he trusted the House would force upon the authorities. (Renewed cheers.) The death of Lilley had excited a very painful feeling in India, the matter had been taken up by the press, and it became clear that it could not be confined to that country. It was at last brought to the notice of the Commander-in-Chief who, in his Memorandum, while censuring Colonel Crawley, had, with the most incomprehensible leniency, left him in a position for which he showed himself so manifestly unfit. ('Hear, Hear'.)

Let the House mark the varying measures of justice, according to the rank of the offender. Paymaster Smales, for writing a letter which was considered insubordinate was brought to a court-martial, tried and cashiered; while the Commanding Officer, acting with the greater injustice and oppression was visited with no severer punishment than a simple reprimand. ('Hear, Hear'.)

But he could not believe that in the face of such facts as had been brought forward, this strange leniency would be persisted in. He hoped that Colonel Crawley would not be left in power, as he had shown he had the will to inflict tyranny and persecution on those beneath him.

Before he sat down he wished to meet two objections. Firstly that he was attacking a man in his absence. The charges against Colonel Crawley were in the possession of the War Office, were widely circulated in the Press and widely believed. It was the natural feelings aroused in him by the events that had determined him to raise the matter.

The second was that the House should not interfere with the ordinary administration of justice. In that, as a general rule, he entirely concurred. But when injustice had been done and redress sought in vain from the ordinary tribunals, it was not only warranted but imperative to interfere.

A final appeal was, therefore, now made to that public opinion, of which the House was the highest representative, and it was in the full belief that this appeal would not be made in vain that had induced him to bring this matter before the House of Commons.

The next to speak was William Coningham, MP for Brighton,[11] who proceeded to make an outright attack on the Duke of Cambridge for his feeble handling of the affair. Quoting from the now famous Memorandum the part where H.R.H. had been regretfully forced to disagree with the officers in high command, and had 'only one course to pursue', he asked, 'And what would the House believe was that course?—to retain the Colonel in the command of the distinguished regiment to which he was such a disgrace?'

It should be remembered that Colonel Crawley had placed these men under arrest for conspiracy, for which there was not a shadow of evidence, imprisoned them without trial and tried to attribute the death of R.S.M. Lilley to excess—and what were the reactions of H.R.H. to all this—he 'could not speak in too strong terms'.

If anything could sap the discipline of the British Army and lower the high spirit of the British soldier, it would be conduct such as this, should it receive the sanction of the House. He therefore hoped the noble lord, in his reply would not confine himself to the conduct of Colonel Crawley, but justify, if he could, the Commander-in-Chief, who, by the issuing of this Memorandum, had made himself responsible for every action Colonel Crawley had committed.

Lord Hartington*, the Under Secretary for War, apportioned the blame for the illegal arrest of the Sergeant-Majors as follows: Colonel Crawley thought he had sufficient evidence to prove conspiracy against them and ordered their open arrest. Major General Farrell was wrong in ordering their close arrest and still more so in obeying the letter and not the spirit of the instruction he had received from Sir William Mansfield.

* The future Duke of Devonshire.

8

Sir William was wrong in ordering the continued arrest, 'in the interests of discipline', and wrong in thinking in doing so he was acting as a police magistrate when remanding a prisoner.

It had next to be considered what could possibly be done by the home authorities in the matter.

If Colonel Crawley was covered by the command of his superior officer it would be perfectly useless to try him by court-martial. Major General Farrell also appeared to be covered technically by the authority of his superior officer for what he had done. But it would be perfectly in the power of the Commander-in-Chief, in conjunction with the Secretary of State, to reprimand these officers in such terms of reprehension as might seem advisable.

But it was incumbent on His Royal Highness to act with the greatest care and caution when dealing with so distinguished an officer as Sir William Mansfield. ('Hear, Hear'.) He was in no way mixed up with the squabbles and petty jealousies that had distracted the 6th Dragoons; though wrong he had acted in what he supposed the best interests of the Army.

As it appeared that the close confinement of Sergeant-Major Lilley was illegal and as his death was probably caused by his confinement, it appeared right to the Government that some reparation should be made to his surviving relatives who were persons in poor circumstances. (Ironical cheers.) The Government had determined to make them such a grant as would be about equal in amount to the pension to which the deceased would have been entitled had he received his discharge at the time of his death. ('Hear, Hear' and a laugh.)

These were the main facts of the case. There were, however, certain collateral circumstances on which further enquiries should be made. There was a marked discrepancy between the statements of Colonel Crawley, as made in his address to the Court-Martial, and the allegations of Lieutenant FitzSimon, the Adjutant, who entirely denied that it was he and not Colonel Crawley himself who ordered the arrest to be made, and by whom the confinement of Sergeant-Major Lilley and his wife was rendered so offensive.

When Colonel Crawley affirmed one thing and Lieutenant FitzSimon affirmed exactly the reverse, it was obviously impossible for anyone at home to come to any decision on the matter until it had been thoroughly investigated on the spot.

Mr Coningham. What about the Memorandum?

Lord Hartington replied that His Royal Highness had condemned in strong terms the conduct of Colonel Crawley but at the time of the Memorandum he was not in possession of the Judge Advocate's opinion or of all the facts that had now appeared. If any moral guilt rested on Colonel Crawley he could only be punished on the point that was being investigated.

It would have been a great deal more satisfactory if he could have come down tonight and handed over to the House a victim who should atone for the wrongs of Sergeant-Major Lilley and his comrades. But he had told them without reserve the facts of this most painful case and he was sure the House would agree with him that even the horror which such a tragedy inspired, or even the disgust which was caused by a contemplation of such mismanagement and such military scandals, would not justify the Commander-in-Chief or the Secretary of State in departing one hair's breadth from the strict rules and principles of justice. (Hear, Hear.)

Mr Alderman Sidney was astonished that the C.-in-C. having expressed his reprobation of this deplorable case, should have allowed Colonel Crawley to remain at the head of the Regiment one day. If the outrage to which Sergeant-Major Lilley had been subjected were committed by civilians, no mere rank would have saved the guilty party from an immediate investigation.

It was not satisfactory to be told that Colonel Crawley was continued in his command simply because he enjoyed the confidence of his superior officers. Why, according to the statement of the noble marquis, that superior officer was himself implicated in the affair . . . the people of England would not be satisfied with the apologetic statement of the noble marquis. Something more must be done; and he trusted the Government would give a direct assurance that those guilty of so flagrant a crime—not to speak of the indecency with which poor Mrs Lilley was treated— would be brought to justice.

Colonel Bartellot thought that the speech of the noble lord would be read in India with some displeasure because the noble lord had endeavoured to shift the blame from one to another. He had implicated those general officers in India who had distinguished themselves on all occasions, and who, he had hoped, would have received from the noble lord complete exoneration in the matter. He had also hoped that the noble lord would have been able to completely exonerate Colonel Crawley. (Oh, Oh.) He was not acquainted with the officer, and only defended him as he would defend any other absent gentleman who was unjustly attacked.

Many one-sided statements had been made in the country; but from a sense of duty to officers in higher command than himself, Colonel Crawley had refrained from answering the statements which appeared in the papers. With the exception of the noble lord, every speaker had denounced him. But India was a distant country and there were many circumstances attending the Court-Martial at Mhow which they had not been able to arrive at.

These Sergeant-Majors were certainly believed to be guilty by Colonel Crawley and General Farrell; and when the evidence and the charge were forwarded to Sir William Mansfield, he said that under the peculiar circumstances of the case, these men were to be kept in close arrest until the termination of the court-martial. If that were so, surely Colonel Crawley had some reason for what he did, and neither he nor any other officer ought to be hastily judged. Men in the Army were under a different rule from civilians and had not the same ready methods of redress which civilians enjoyed. A man who was serving his country abroad ought not to be assailed in his absence unless sufficient reasons for doing so existed. On previous occasions Colonel Crawley had shown great kindness to the R.S.M. He (Colonel Bartellot) was not going to blacken the memory of the Sergeant-Major. All he wished was to show that Colonel Crawley was not answerable for his death. The day before it occurred Colonel Crawley had asked General Farrell whether the close arrest might not be changed, so that at the time Sergeant-Major died he was in ordinary arrest.

With regard to the adjutant, Lieutenant FitzSimon, it did not appear from the remarks of Sir Hugh Rose that that officer stood very high in his opinion. The Commander-in-Chief of India was not likely to be prejudiced or have any bias upon such a matter and he had said that the annoyance caused by setting a sentinal in the quarters of Sergeant-Major Lilley might be traced to the adjutant's neglect of duty.

(Mr Coningham—That was the statement of Colonel Crawley himself.)

General Farrell, who was in this country said, in a letter which he (Bartellot) held in his hand:

'You will perceive that another letter has appeared in *The Times* this morning, signed "Civilian", wherein a gross misrepresentation of the truth has been made in regard to Lieutenant FitzSimon. The case is as follows—Lieutenant FitzSimon wrote commenting, as I conceive, in a disrespectful manner on Sir Hugh Rose's order, and this letter I advised him privately to withdraw. He said he would do so on my order. "No, no," I said, "you can send it if you like. If you withdraw it, it must be your own act and deed," and as his own act and deed he did withdraw it. If I recall aright, my assistant adjutant general was present.

'With reference to Colonel Crawley's dropping the charges of conspiracy against the sergeant-majors, he had nothing to do with it. I, as general commanding the division, received orders from the Commander-in-Chief, directing them to be forgiven, and a warning order to be read out to all the non-commissioned officers assembled for the purpose. This I duly read, and sergeant-majors Wakefield and Duval were released accordingly. A summary of evidence was sent in by Colonel Crawley to me against the sergeants which I forwarded to the Commander-in-Chief, and it was his order that no prosecution against them should take place, and that they should be forgiven.'

This showed that Colonel Crawley was not to blame and he (Colonel Bartellot) was sure that the sense of justice entertained by Englishmen would lead them to feel that this was not a case to be decided in a hurry upon any one-sided view. The sick wife of Sergeant-Major Lilley died, and he was sure that no one deplored the loss more than Colonel Crawley. But Colonel Crawley's wife was now in this country, and anonymous letters had been sent to her, declaring that on her husband's arrival in this country, a charge of manslaughter would be preferred against him. Everyone must deprecate such conduct.

The Commander-in-Chief had retained Colonel Crawley in

command of the regiment. This was prima facie evidence that he was not guilty(!)

In conclusion he would appeal to the House and the country not, for the sake of doing what was alleged to be an act of justice to a man who had died in the service, to do an act of injustice to a man who was still serving his Queen and country to the best of his ability.

Mr Sergeant Piggot (Reading), was surprised that the hon. and gallant gentleman should have charged the noble lord with not having done his duty towards these officers, for they were all acquitted by the noble lord of any share of blame. The hon. and gallant gentleman had told the House that Colonel Crawley was not to blame; but he wished the hon. and gallant gentleman had gone one step further and informed the House if Colonel Crawley was not to blame, who was?

Captain Archdall, as an old Inniskilling officer, said he felt for the character of the Regiment. He thought that when the discipline and good conduct of a regiment declined to such a state as the Inniskillings had done, this might well be the cause for removing the Commanding Officer.

Mr Headlam (the Judge Advocate General), remarked that Colonel Crawley was not removed because what he did was sanctioned by his superiors. After that fact was established, it was impossible to remove him.

16

The Press in full cry

Following the Commons Debate, the pressure of protest that had been steadily mounting over the weeks, erupted in the Press. In the words of the *Grantham Journal*, 'The seriousness of the business may be judged when all the leading press are on the same side crying for justice in different keys.'

The *Manchester Guardian* handed out congratulations all round—'It is one of the best functions of the press to drag out such cases as this into the full light of publicity. *The Times* never exerted its powers for this purpose in a case in which their exercise was more imperatively called for than in this of Colonel Crawley, which on Friday moved the indignation of the House and is now exciting the strongest feeling throughout the country.'

The Times, in fact had, up to the Debate, been content to let J.O. make the pace through its correspondence columns. It now cast aside all restraint with a leader which was remarkable not only for its whole-hearted condemnation of the military authorities but for the highly emotional language in which it was phrased. On the same day it published another letter from J.O. replying to the defence of General Farrell put up by Colonel Bartellot in the Debate. In view of the parts still to be played by the paper and its correspondent in the affair, it is worth noting the quality of the contributions made by each at this stage.

According to the leader, all the aspersions cast on the memory of Lilley were proved to be false and it was now 'a simple and naked case of a veteran soldier of irreproachable character persecuted to death by the oppression, or connivance at oppression, of officers of high rank.

'. . . It sounds a simple thing to order three men into arrest. To understand, however, what arrest meant in this case, we must go into detail. We must remember that the court-martial was sitting, not in England, but in India in the months of April and May, the most suffocating period of that climate, when the heat is untempered by wind or rain, and when all creation seems to be gasping for the cooling showers yet to come.

'The "Black Hole of Calcutta" has been destroyed, and so has the place of confinement to which these men were consigned. Mr Dudley Fortescue states in his speech in the House of Commons, that it has been pulled down as unfit for the habitation of troops. The rooms were "more like ovens than human habitations", having bomb-proof roofs, which retained, night and day, the heat of the tropical sun. We need not follow the fate of the other two prisoners; for although one of them was found to be a maniac when the order for his release came, yet they did not actually die in arrest.

'To Sergeant-Major Lilley, however, the confinement had special aggravations. His wife was sinking in a mortal disease in the same room, or oven, in which Lilley was confined. She was just kept alive by wine and spirits prescribed by the hospital doctors, and supplied by a native sutler. Of course the poor creature could not move; and anyone who has felt the heat of the tropics can imagine what must have been the scenes presented by the interior of that dwelling, inhabited by two persons, one of whom could not move, and the other of whom was not allowed to go out. The sentry paced up and down outside. But it seems that Colonel Crawley, whose information was accurate, heard that this wretched woman had attempted to communicate with another woman outside . . . he ordered the sentry to pass from the outside to the inside of the cell. It was told to Colonel Crawley that his prisoners were already loathsome in their privacy and that their situation must become too horrible for endurance if a strange man were thrust inside their little den of torture.

'The answer of the Colonel is recorded, and is of a class which history has often thought worthy of note when proceeding from some special tyrant or particularly callous man. The sentry was

posted inside the room. This arrest continued from 28 April to
25 May, and on that day, in stench and in suffocating heat, and
in mental agony, produced by oppression and calumny, Lilley
died . . . Poor Lilley! Was it worth his while to practise all the
soldier's virtues for 20 long years in every climate to be at last
persecuted to his death by his Colonel with the approval of two
very eminent Generals, and then consigned to infamy, and made
to point a moral by the Commander in Chief in India? The men
who heard the moral and knew the facts must have been much
edified.

'But all has come right at last. The generous British Government has recognized the injustice suffered, and has made a com-
pensation by awarding somebody out of the public money, a
sergeant's pension! The Commander-in-Chief has publicly
acknowledged that poor Lilley's memory is clear from the stigma
cast upon it; and he has condemned all the officers concerned in
the tragedy to a gentle censure, a general eulogy, and a complete
immunity!

'Such is the justice dealt out upon four officers for the cruelty
which causes death, or the carelessness that authorizes cruelty.
This will not do. Putting all feelings of right and wrong on one
side, one cannot wipe out this great crime as you would condone
a card-table quarrel. You cannot maintain discipline in this or in
any other army if you proclaim that men may be persecuted to
death without responsibility or punishment attaching to the per-
petrators of the crime.

'No greatness of name, no length of service ought, in such a
case as this, stand between the oppressor and the law. If you
would have the soldier obey, you must make him feel he is protected. Here is an unpunished crime of blood-guiltedness, and the
men who perpetrated this act must be put upon their justification
before some unsuspected tribunal, or law and justice are naught
in India.'

By contrast with this melodramatic outburst, J.O.'s letter was
confined to setting out the evidence that Crawley and Farrell had
used intimidation and lies to suppress information prejudicial to
their case and that their word had been accepted uncritically by
their superiors.

It was essential, he said, for FitzSimon's prospects that he should clear his name of the slurs cast upon it by Crawley, and this he had been able to do. As to the letter which Farrell was said to have persuaded him to withdraw in his own interests—if the Commander-in-Chief would read it, he would find it to be a perfectly respectable and proper letter to be written under the circumstances, and that it did not reflect directly or indirectly on the General Order of Sir Hugh Rose. It merely put in a plea of 'not guilty' and produced evidence to support it. Its suppression by Farrell and Crawley was clearly a grave military offence—the more grave because the letter, in clearing Lieutenant Fitz-Simon, must necessarily incriminate them.

On the suggestion that Crawley had ordered the release of the Sergeant-Majors from close arrest before Lilley died, J.O. wrote '. . . if Colonel Bartellot will produce a copy of the order which he says Colonel Crawley gave him to change Lilley's close arrest into ordinary arrest before he died, its date—if it bears a date—will show that this order was not given before the man was insensible with apoplexy.'

Sir Hugh Rose had passed censures on several individuals for offences that were in no way alluded to in the evidence at the Court-Martial. 'No witness ventured to hint that Sergeant-Major Lilley was other than an extremely sober man; no witness ventured to impute before the court to Lieutenant FitzSimon the responsibility for having kept for many days, sentries close to Mrs. Lilley's sick bed. The animadversions by the C.-in-C. on these points were, therefore, clearly not based on the evidence before the Court-Martial, but on private and ex parte information, supplied to him by some unknown accuser against these two officers, to whom no opportunity of reply or explanation had been afforded before they were thus publicly stigmatized in a General Order, and we now know that these animadversions, as far as Sergeant-Major Lilley was concerned, were utterly unmerited, and that when Lieutenant FitzSimon respectfully attempted to defend himself against them, he was told it would be insubordinate for him to do so, and was intimidated into withdrawing the letter which he had written with that object.'

J.O. concluded that when the Duke of Cambridge decided to continue Colonel Crawley at the head of the 6th Dragoons because the Generals in immediate command expressed their high opinion of him, he must surely have forgotten that these Generals were the very individuals who, by their approval of his conduct, had so sheltered him as to render the punishment of an offence, the gravity of which was fully admitted by the military authorities at home, almost impossible. The commendation bestowed by these officers on Colonel Crawley was merely an attempt at self-preservation on their part.*

Other shades of public opinion may be gauged from a more or less random sample of comments from other dailies and weeklies.

The *Manchester Guardian* wanted to know why Crawley had not yet been dismissed, Farrell relieved of his District Command and Mansfield admonished. On the C.-in-C. India they had this:

'That Sir Hugh Rose should have extended his protection to Colonel Crawley will not much surprise those who have watched Sir Hugh's persistent efforts as Commander-in-Chief in India to destroy the reputation acquired as a General in the Indian Mutiny. Sir Hugh has the misfortune of most elderly ladies' men of believing himself irresistible, and a petticoat is sure to be mixed up in any delicate or difficult affair in which Sir Hugh comes before the public. The petticoat enters into social scandals and rows in India to a degree hardly conceivable in this country.'†

The *Grantham Journal* also criticized the conduct of Sir Hugh who 'has pretty well worn out the patience of the Army in India since he has been Commander-in-Chief. His military services are forgotten in the multitude of annoyances he is daily inflicting on those he does not like; and with such a temper as his, there are many he does not like. His patronage of Colonel Crawley is likely to shorten the term of his command.'

The paper's London Correspondent continued: '. . . it appears, in India, that an officer may be ruined for writing a true but imprudent letter to his Colonel, but that Colonel, if he manages well, may kill off any inconvenient officer by illegally imprisoning him until he dies or becomes insane from heat, want of exercise and vexation.'

---

*J.O. also referred scathingly to the address by Crawley to the Regiment after Sir Hugh's General Order had been read out.
† Sir Hugh was a lifelong bachelor.

Concern was also expressed as to the effect Lilley's death was likely to have on recruiting.

The *Lincolnshire Chronicle*—'Such a mode of administering justice in the Army, unless visited by an ample and adequate condemnation, will inevitably tend to damage the popularity of the service among the class of people from which its rank and file are recruited, especially in this County, where the case of poor Lilley has naturally excited peculiar indignation from the fact that he was a Lincolnshire man.'

The *Spectator* felt even more strongly—'. . . The issue at stake is nothing less than the popularity of the army with the classes which furnish recruits. These classes forget nothing of individual interest, and five battles may be lost with less effects on their minds than one clear case of fatal but unpunished tyranny.'

The *Nottingham Review* however was more optimistic and ended its comments with a rallying call. 'Fortunately for the help of the Army against tyranny, a large portion of our citizens have banded themselves together in our Voluntary Corps, and they will feel the case come closer to their hearts and heads than ever before. Let them come to the assistance of their brothers in arms, those who stand before them in the first line of defence, and insist that the punishment for this outrage shall be such as to prevent its recurrence hereafter.'

But while the demand for someone to be punished was universal, the practical need to balance justice and expediency was not entirely forgotten. For if justice was to be the sole criterion, punishment ought to fall on all the high-ranking officers who had become enmeshed in the quarrels at Mhow.

The *Spectator* rationalized the problem by pointing out that although the train of responsibility went back to General Mansfield, 'to ruin, in a military sense, one of the ablest officers in the service—the man to whom the country would look in the event of any war in Asia or Eastern Europe, would not be expedient. So to ruin him in order that Colonel Crawley, the officer really responsible might escape the heavier charge—for that must be the effect of punishing General Mansfield—would be simply a folly.'

A possible way out was suggested by the *Lincolnshire Chronicle*. Crawley should not be immune from punishment simply because two of his superior officers had sanctioned the arrests, for they had clearly 'acted on a misapprehension'.

A less sophisticated and more representative view was contained in a letter addressed to the Duke of Cambridge at the Horse Guards.

> 384, High Street,
> Cheltenham.
> 8 June, 1863.

His Royal Highness,

Will you do me the honour (being too poor to obtain legal assistance) whether we, the friends of poor Sergeant-Major Lilley, can have Colonel Crawley arrested on his arrival in England on a charge of manslaughter. I cannot allow myself to express my feelings on the conduct of that brutal fellow, lest I should overstep the bounds of courtesy due to your Royal Highness.

> Yours, &c.,
> C. Pembruge Langston

On 15 June, when the Duke went to the Lords to explain his position as C.-in-C. in relation to the whole affair, he heard Lord Shaftesbury refer to the treatment of Lilley by those in authority over him as 'so frightful a picture as had never yet been brought under the notice of the British public', and ask whether the Government intended to institute further inquiries.

The Duke explained that the Army in India was entirely under the control of the C.-in-C. there and that his role was that of a referee to whom a last appeal could be made. When he drew up his Memorandum on the Mhow Court-Matrial he had not been in possession of all the facts, and only in the last ten days had he been in a position to advise that Colonel Crawley should be brought to a court-martial. Explaining the long delay in arriving at this decision he said it took at least three months before a reply could be received to any question sent to India, and further delay arose because of the time taken between Sir William Mansfield at Bombay and Sir Hugh Rose whose H.Q. was in Bengal.

He believed that both these officers had committed a serious error of judgement but that neither had any desire to be cruel or unjust. He ended by strongly denying that he had acted because pressure had been put on him.

The Secretary for War (Lord de Grey and Ripon), who appears to have been a man of few words, remarked that he need say no more as the C.-in-C. had intimated he would bring Colonel Crawley to trial.

Having heard the news they had been waiting for the Press now addressed itself to the question of where the Court-Martial ought to take place. On this as on the other issues connected with the affair, J.O. continued to lead forcefully and coherently. *The Times* on 18 June published a letter in which he gave his reasons why it must be held in England.

First, many of the material witnesses—Farrell, Smales, Dr Barnett, FitzSimon and Wakefield—were in this country. Several were not likely to return to India if they could help it, nor could the authorities compel them to do so.

Second, were it to take place in India, Sir Hugh Rose, who had already prejudged the case, would have to select the president of the Court-Martial on Colonel Crawley. He had known all the relevant facts about the imprisonment of the R.S.M. but had not thought it worth while commenting on them in the General Order in which he 'eulogized Colonel Crawley as few commanding officers had been eulogized by a Commander-in-Chief'.* If Crawley failed to clear himself, Sir Hugh would himself be inculpated by the fact that he had allowed the arrests to pass without censure. And if Crawley did clear himself he could only do so by shifting the responsibility for what he had done on to the shoulders of those who had sanctioned it.

'In either case,' commented J.O., 'a mandate from the Horse Guards, ordering Sir Hugh to bring Colonel Crawley to a court-martial for conduct known to and approved by himself, will partake of the nature of the bowstring which used formerly to be transmitted by the Sublime Porte to peccant Passhas for purposes of self-strangulation.'

The Spectator too thought Sir Hugh Rose had so deeply implicated himself by his remarks on the death of Lilley that there could be no confidence in the impartiality of any court he might assemble to try Colonel Crawley. It also believed there would be an excessive and just reluctance of Indian officers to convict a commanding officer on a charge which, however worded, would

* From Crawley's address to the Regiment after reading the Order.

be virtually one of tyranny. Senior officers remembered that it
was by the home Government's interference with the powers of
commanding officers that discipline had been weakened in the
old native army.* They were by training inclined to suspect any
move that might repeat the blunder. The best solution would be
to call the Regiment home and hold the trial in England.

Before recounting the steps actually taken by the Authorities
it is important to note that, after the first outburst of public
indignation following the Commons Debate, signs of a reaction
began to appear.

In the *Lincolnshire Chronicle*'s 'London Talk' came this:
'The Exhibition building has been the most exciting topic at
home; next to it comes the Crawley case which has been so
singularly misrepresented by the daily newspapers.'

The misrepresentation this writer objected to was the criticism of the Duke of Cambridge for failing to take action over the
death of Lilley months before. This, he thought, was unjust, for
the Duke had only recently received much of the information
relative to the case. The Duke was still regarded by the public
and those best able to judge, as 'the soldier's best friend and a
straightforward gentleman'.

The Spectator, after a fortnight's digestion of the available
information, decided that too much sympathy was probably being
wasted on Mrs Lilley. She 'was dying of diarrhoea, and an
Indian bedroom has, and can have, no curtains but net; but so
infamous was the system of Indian barrack structure till 1859,
that Mrs Lilley must have slept for years under surveillance
hardly less close than that which killed her husband.'

In the light of future events it was rather ironical that
Crawley's defenders in the Press preferred to remain unidenti-
fied while his opponents like Smales and J.O. wrote openly. On
18 June, *The Times*, steadying itself after the outburst of ten
days before, published a letter from *Audi Alteram Partem* expressing the 'relief of Colonel Crawley's friends that a Court-
Martial, which will practically involve an inquiry into the
conduct of all concerned' had been ordered.

He concluded: 'In the meantime, you may rest assured that
Colonel Crawley is able both morally and legally to meet the

* This was sometimes advanced as one of the main causes of the Mutiny.

charges that may be brought against him; but he entreats that he may not be convicted before he is tried, nor condemned before he is convicted. No one deplores poor Lilley's death more deeply than Colonel Crawley.'

17

A kind of pardon

By a strange irony, Smales was almost forgotten in the present furore. Yet, for a while he was no doubt content. For without the unforeseen tragedy, all his clamour for justice would have been ignored by the Military Authorities, whereas now there was the hope that his fortunes would advance on the tide of public sympathy for the dead man.

After the Judge Advocate General had been forced to declare the arrest of the Sergeant Majors illegal, and since they were at the time Defence witnesses, the question was bound eventually to be asked how far that illegality had affected the validity of the Trial itself. On 9 June the same legal expert produced the answer to that question.

He began by adhering to his original view of the previous November, that Smales, in spite of the objectionable features of the Trial, must be considered guilty of writing an insubordinate letter. But as the Court Proceedings had been taken up with discussing the alleged false and malicious statements in the letter and as Colonel Crawley had been given every facility to prepare his case including unlimited access to Defence witnesses as well as his own, Paymaster Smales should also have been allowed to communicate freely with his witnesses.

Colonel Crawley had not only arrested some Defence witnesses, so preventing the Paymaster from communicating with them, but had misled the Court by alleging that they were untrustworthy from having engaged in a conspiracy against him. For this reason the validity of the whole Proceedings was

129

affected and the sentence of cashiering passed on the Prisoner could not be executed without straining the law. Her Majesty should, therefore, be recommended to grant Mr Smales a free pardon. However, as Mr Smales had admitted writing the letter, the pardon need not necessarily involve restoration to full pay.

This decision naturally left the Authorities in as great a quandary as ever on what to do about the ex-Paymaster. Two days later they informed him that some time must elapse before they could let him know what had been decided.

Meanwhile Smales had at last found an advocate for his cause. On 12 June, ignoring an appeal by Palmerston to refrain, William Coningham, the radical MP for Brighton, again raised the subject of the Mhow Court-Martial on the grounds that the case of Lilley was only part of a much larger question. The Under-Secretary for War, he said, had regretted being unable to produce a 'victim' for the House, but now that he had read the proceedings of the Court-Martial, he was not satisfied with being presented with a 'scapegoat'. Colonel Crawley was not the only officer compromised. Two general officers and the C.-in-C. India were implicated as well as the C.-in-C. of the British Army at the Horse Guards.

At this point he was interrupted by General Peel on a point of order. But the Speaker having ruled in his favour, Coningham commented that the interruption was entirely in accordance with the whole conduct of the military authorities. They were all for suppression; but so long as he had the honour of a seat in the House he would not allow so monstrous an outrage to be perpetrated without protesting. He demanded that justice should be done, even if those to whom blame attached were general officers and included the Commander-in-Chief of the Army. (Oh! Oh!)

It had been said that the House of Commons was a house of Colonels, and one would think from the groans of those who sat opposite that their sympathies were enlisted on the side of the men who had deliberately and in cold blood, acted in the manner to which he was about to advert.

The question was one in which he thought the House was bound to interfere. He might be told it was merely a question of discipline and that the poor Paymaster might be crushed with impunity; but he hoped to satisfy the House that the justice of the case called for a rigid and searching investigation of the distinguished officers concerned. He would summon His Royal Highness himself to the bar of a tribunal where he would be judged, not perhaps by his equals but by those who would not stand by and see injustice done. The Commander-in-Chief, with the strangest weakness, had altered his first decision and had determined that Colonel Crawley should be brought before a court-martial. He had by that act condemned himself.

Coningham then proceeded (with frequent interruptions), to outline the case for an investigation into the Court-Martial as it affected Smales. That he received no support may not have been entirely due to the unpopularity of his cause. For he had a failing not uncommon in those quick to respond to any injustice. When others failed to share his indignation, he turned his invective upon them. His manner of speaking, nervous and snappy, was also against him.

The Government spokesmen put up the classic defence against the outsider whose logic is hard to refute. Lord Hartington defended the Duke of Cambridge by saying that although the Speaker had decided he was in order, there could hardly be any difference of opinion as to the hon Member's *good taste*.

By the end of July, realizing that a 'wait and see' policy was getting him nowhere, Smales jogged the Horse Guards by asking whether his sentence had been remitted. Three days later— a full seven weeks after the Judge Advocate General had recommended it, he was informed that the Queen had pardoned him.

Press reports of his reprieve were approving and though the decision not to reappoint him to the Inniskillings was agreed to be reasonable, it was generally forecast that he would be offered the paymastership in another regiment in due course.

But when three more weeks passed with no further information as to his future, Smales sent in a request for the issue of his back pay. He received a reply, brief and to the point:

'. . . I am directed to acquaint you that the Secretary of State for War has been advised that the pardon which her Majesty has been pleased to grant you has left you in the position of a civilian. I am therefore to inform you that you are not entitled to receive any pay, and, further that it is not the intention of Earl de Grey and Ripon to submit your name to the Queen for appointment as a paymaster.

Ed. Lugard.

While *The Times* had itself no direct comment to make on this surprising interpretation of the term 'pardon', it published an extract from the *Express* on the subject:

Everyone knew what 'consider yourself horse-whipped' meant. But in the case of Mr Smales the War Office had invented a new formula—'Consider yourself pardoned'. For, after having been illegally cashiered, the unfortunate Paymaster was to consider himself pardoned, yet was to remain under precisely the same penalties as if he had been legally cashiered and had received no pardon.

If Mr Smales was guilty of any offence that deserved dismissal from the Army, let him be tried again or let him be told the grounds on which he was being excluded.

A reply to this came from an anonymous correspondent explaining why he thought Smales should not be re-employed.

'. . . A Paymaster has large sums of public money intrusted to his charge and the appointment can, therefore, only be given to officers in whom the authorities have the fullest confidence.

'It appears from the newspaper reports of the proceedings at Mhow that, when Mr Smales' accounts were examined by the appointed authorities, there was a deficiency of public money not far short of £2,000. Therefore, before giving the sympathy which is asked for Mr Smales, it would be desirable to know whether there is any foundation for this statement.'

The next day Smales again vehemently defended himself in a letter in which he described his efforts to clear himself of the accusations against him. He said he had repeatedly written to the Horse Guards, the War Office and the India Office complaining about the way he, as a financial officer, had been treated by the Indian Military Authorities. Two of the departments had never answered his letters while the third, the War Office, had referred him for 'redress' to one of the others.

'. . . no later than yesterday', he wrote, 'I was compelled to waste the whole of my day backward and forward to the War Office, endeavouring to get an interview with Sir Edward Lugard on the very subject now referred to, and after notifying

to his private secretary the nature of my visit. On my final visit
at 5 o'clock in the evening, after waiting half an hour, I was
coolly informed that Sir Edward Lugard 'really could not see
me on any such subject'.

Here, with ex-Paymaster Smales still fighting for 'redress',
and the hour approaching when his one-time Commanding
Officer would be standing trial, it is time to bring up to date the
course of events as they affected some others in the affair.

18

Sir Hugh Rose remains adamant

On 18 June, a letter was sent from the Horse Guards to Sir William Mansfield expressing the views of the Duke of Cambridge on the arrest of the Sergeant-Majors in the light of all that had happened in the past weeks. Its composition must have called for a good deal of care.

After pointing out that he had, from the first, disapproved of the arrest and long confinement of the N.C.O.s, the Duke then referred to the opinion of the Judge Advocate General which he enclosed for Sir William's perusal. This, he said, was so plainly stated and so clearly reasoned that he only felt it necessary to call Sir William's attention to it, and express his deep regret that an officer of his high standing and reputation in the Army, and occupying so prominent position, should have been led into so great an error of judgement as to have sanctioned the continued arrest of these non-commissioned officers without sufficient grounds being shown for bringing them to trial.

He knew that Sir William was only anxious to maintain discipline with justice and that the unfortunate course he had adopted was solely due to 'an error of judgement'. He was convinced that had Sir William known the manner in which the close arrest was carried out he would have been the first to condemn it.

He could not, however, give the same immunity to Major-General Farrell and Lieutenant-Colonel Crawley who were on the spot and responsible for carrying out the confinement. He had therefore decided to have the conditions of the close arrest investigated by a Court-Martial on Colonel Crawley, and desired Sir William to convey to Major-General Farrell 'the marked disapprobation' with which he viewed his conduct.

Farrell, now living in retirement at Winchester and for whom life's battles were almost over, could probably accept any reprimands with philosphical indifference—even one from the Duke who refused to believe he had advised FitzSimon to withdraw his letter 'as a friend'. But the high-ranking officers still in India were by no means content to take lying down the severe criticisms directed at them from so many quarters.

In public they were precluded from defending themselves and had to rely on influential friends to salvage their reputations. Behind the scenes both Sir William Mansfield and Sir Hugh Rose not only protested but did all they could to control the course of events.

The former politely refused even to admit to errors of judgement in his decisions on the arrest of the Sergeant-Majors and, supported by his military legal adviser, tried to disprove the arguments of the Judge Advocate General.

But it was Sir Hugh Rose, who carried the ultimate responsibility and had become so personally involved in the affairs of the Inniskillings, who appeared most aggrieved at the turn of events. Over many months he had expended a vast amount of time and effort attempting to prove the justice of his contentious remarks on the Mhow Court-Martial. Apart from several long letters refuting the criticisms of himself made in Parliament and specifically in *The Times*, he set about compiling a detailed History of the Inniskilling Dragoons from their departure for India in 1858 to the Court-Martial on Smales, four years later.

This highly personal narrative sheds a good deal of light on events but reveals even more of Sir Hugh's mind and the way his subordinates pulled loyally with him. He develops his arguments at great length. Yet because his premises are doubtful he generally fails to convince. To defend his reputation he not only has to present Crawley as the innocent victim of sinister plots and circumstances over which he has no control, but to accept that individual's assessment of the characters of all who had contributed to the dangerous situation in which he now found himself.

That the home authorities had disagreed with his views from the start was shown by the Duke of Cambridge's Memorandum of 18 December in which he implied that Crawley was himself the main cause of the trouble. Equally important, he had firmly dissociated himself from Sir Hugh's strictures on the dead R.S.M and on this the two Commanders had now become completely irreconcilable.

Correspondence between them in the Spring of 1863 shows the Duke holding to the view that there was no real evidence that Lilley hastened his death by drinking and that his previous life had been marked by good conduct and sobriety. Sir Hugh continued to maintain the opposite and promised he would be sending 'proof' that Lilley was not at all the superior N.C.O. he was made out to be.

The 'chain of evidence' which eventually arrived purporting to prove this contention may be summarized as follows.

Lilley drank an undue quantity of spirits at the time his wife was dying, 'when of all others, he should have refrained from such culpable indulgence'.

He was utterly unfit for the post of R.S.M. as proved by the fact that Sir William Mansfield considered him so and ordered him to be removed from it.

Colonel Crawley had been compelled on one occasion to tell Lilley he was unfit for his duty. (This was obviously deduced from the R.S.M.'s own evidence at the Mhow Court-Martial.)

Pointing out that Lilley was responsible for the Regiment's discipline, Sir Hugh quoted the case of the N.C.O. who tried to 'screen' a drunken man for guard duty on the chance he would pass the R.S.M.'s inspection but concluded somewhat lamely— 'which, however, he did not do on that occasion'.

If Lilley and the other Sergeant-Majors had not been arrested, they might have circulated Smales' insubordinate defence amongst the soldiers and so gained their support against their C.O.

Lilley used 'the most wicked and disgraceful epithet which the English language contains' against his C.O. This was 'proved' by Moreton's statement and 'almost admitted' by Lilley himself.

The R.S.M. had instituted a secret and illegal 'arrack' (native drink) fund as soon as the Regiment arrived in India from which he derived huge profits. Its existence only came to light following an investigation after his death, so that he must have carefully concealed it from his two C.O.s. The presence of this fund explained the large bill for spirits during his arrest which he could not possibly have paid for out of his Army pay.

A committee of sergeants convened for the purpose discovered that Lilley had at his death 2,000 Rupees (£200) standing to his credit and unaccounted for. Shortly before he died he was seen by a sentry to destroy an account book and was heard to say, 'Thank God! there is an end of the arrack fund.'

Old soldiers, wrote Sir Hugh, are adept at hiding their passion for drink. They drank 'copiously but steadily', but abstained from drunkenness.

So it was with Lilley, who was a man of very intemperate habits but had managed to hide them from his two vigilant C.O.s, Colonels Shute and Crawley.

He had just received remarkable proof from Major-General Green (Farrell's replacement), that even at the time of his death Lilley was unable to resist his craving for drink. The two sentries over him stated that 'he drank off a tumbler of brandy, shortly after which he began to talk loudly and wildly to his wife, who was unable to quiet him; that he talked in the same way to the wife of another Sergeant-Major who came to assist; that the manner of his talking showed that he was wandering in his mind; that he jumped out of bed and struggled hard with the sentry who endeavoured to put him back; that after he was quieted, Mrs Lilley begged the sentry to give him another half tumbler of brandy, a teaspoonful at a time; that his teeth were clenched, so that there was difficulty in giving him the brandy; that he began breathing very heavily, and that, after relief came he died.'

In describing the quarters where Lilley was confined, Sir Hugh felt obliged to refute assertions that their inadequacies contributed to his death, yet had to admit that the Regiment was housed in native cavalry stables converted into barracks.

He was further embarrassed by some comments by Farrell in his half-yearly Inspection Report on the Inniskillings in November, 1862, which commented on the 'lamentable amount of sickness and mortality in the whole Mhow division'. This was generally due to malaria. But the Inniskillings suffered in addition from bad accommodation, 'old Bengal Cavalry Lines', in which they were housed.

This statement evidently conflicted with the impression Sir Hugh was hoping to create. He quoted Sir William Mansfield as saying that Farrell's unfavourable report was to be regretted. The Inniskillings had suffered no worse in the epidemics than the 72nd Foot, which was housed in a magnificent range of new barracks.

But having 'proved' the old buildings were not injurious to health, Sir Hugh went on, paradoxically, to say he had written to the Indian Government, in the name of humanity, urgently requesting more money for the construction of new barracks. This, he contended, proved that the military authorities were in no way responsible for the accommodation at Mhow.

As to the quarters occupied by Lilley, Major Champion, the Assistant Adjutant-General at Mhow, had sent him precise information that the first building was a three-roomed, staff-sergeant's bungalow, and better than most quarters allocated to a sergeant-major. Sir Hugh, who, incidentally, had not seen the buildings added, irrelevantly, that in the Mutiny campaigns officers from generals down would have been glad of such quarters.

The second quarters were admittedly inferior to the first, yet still consisted of two rooms and a verandah. The R.S.M. had been removed because his first bungalow had to be pulled down to make room for the new barracks now under construction.

Considering that the Horse Guards were in possession of all the documents bearing on the subject of Sir Hugh's protests, some of his remarks appear naïve to say the least.

On the criticisms aroused by the treatment of the Sergeant-Majors, he wrote that the benevolent system of the British Army which had for its object the welfare in every respect of the British soldiers, provided ample means of redress for the complaints, however slight, the humblest of them.

He blamed Dr Barnett, who attended Mrs Lilley, for not reporting anything he may have thought improper in the conditions of her husband's arrest. It was his bounden duty to do so, since Military Surgeons were responsible for the health of all their patients, and 'their authority in that respect might be said to be supreme'. To prove his point, he then cited the rather unfortunate example of flogging a soldier, at which a surgeon had to be present to pronounce whether he was physically capable of receiving his full quota of lashes (still a part of the British Army's 'benevolent system').

Of the many inconsistencies in Sir Hugh's 'History', his comments on Colonel Shute deserve notice. On the one hand he accuses him of favouring officers of social standing; on the other he uses for quite a different purpose that officer's letter to the Duke of Cambridge (back in 1858), in which he complained that Weir was a trouble-maker and had been 'an indifferent R.S.M. and a very moderate adjutant'.

From it he infers that successive Inniskilling C.O.s had failed to take enough care in selecting their N.C.O.s. For, with 'peace, good quarters and a popular service' they ought to have been able to train the right type of man. The public had recently complained that not enough deserving N.C.O.s were being commissioned. But when, as in Weir's case, a wrong choice was made, 'neither the officer so raised, nor the officers of a *superior class in life* with whom he associates, nor the soldiers he has to govern, are benefited by what is a disadvantageous anomaly, the transfer of a man of a very different, inferior and less educated class in life to one superior in all these respects.'

Now these are the very sentiments one might have expected Sir Hugh to roundly condemn, and which no doubt he would have condemned as snobbery, had they been expressed by Shute. Yet, by a strange reversal, Sir Hugh proceeds to use them to accuse that officer of repeating the mistake of his predecessors—that is, recommending Lilley for a commission; a man who had since been reduced from R.S.M. for misconduct and since his death was 'proved to have been guilty of intemperance and bad conduct'.

In spite of his readiness to accuse others of snobbery, Sir Hugh was not averse to claiming for himself immunity from criticism because of his social standing. Complaining of the accusations that the Military Authorities in India had failed to prevent the cruel treatment of their subordinates, he said it might have been hoped 'their position, education, character and antecedents' would have protected them against such imputations.

He was particularly pained at the suggestion by the Earl of Shaftesbury that Crawley's trial must take place in England because the case had already been prejudged in India by the fact of Lilley's death having been attributed there to apoplexy caused by habitual intoxication. On this he wrote: 'I have the honour to know personally the Earl of Shaftesbury, and should have hoped that his knowledge of me, as a gentleman by birth, education and feeling, would have freed me from such a serious imputation at his hands.'

The Duke had been placed in an extremely delicate situation by Sir Hugh's Remarks on the Mhow Court-Martial which he could neither condone nor condemn in public. And though he had at first managed to smooth things over, once the affair became public, his attempts at diplomacy had laid him open to the charge of weakness. Not surprisingly, therefore, from then on the disagreement between the two was carried on in private.

Sir Hugh requested more than once, as a matter of justice, for his defence to be published. But this was never acceded to as such a move would inevitably inflame public opinion and force the Horse Guards to state unequivocally their disagreement with the India Command. Even for his own reputation Sir Hugh's so-called 'proofs' could not be allowed to fall into the hands of the critics of the calibre of J.O. Understandably then, whenever he thought it necessary, the Duke quietly overrode Sir Hugh's decisions without actually undermining his authority.

Immediately following the Commons Debate in June, he des-patched two letters to India bearing on the now notorious 'Remarks' on certain Inniskillings' officers. Swindley and Weir, unlike the unfortunate FitzSimon, had succeeded in having their letters of protest against the accusations made by Crawley in his Closing Speech forwarded through the official channels.

A good deal of correspondence had followed, culminating in Sir Hugh's reluctantly accepting Weir's explanation that he was unaware of the evidence Lieutenant Bennitt was going to give and so could not have warned him beforehand of the 'mistake' he made in confusing the Watering Parade with the Muster Parade. In the case of Swindley, however, his explanations regarding the destruction of the wine-books and his alleged refusal to lend Crawley the eating utensils from the Officers' Mess at Ahmed-nuggur were rejected.

But the Duke now saw things differently. He informed Sir Hugh that he accepted the explanations offered by Captain Weir and was glad to find the General appeared to take the same view. But on Major Swindley he felt compelled with great reluctance to differ from the opinion expressed by Sir Hugh and bound to accept the explanation offered.

Regarding a statement that Swindley was pursuing a line of conduct tending to keep alive the bad state of things that had latterly existed in the 6th Dragoons, he was perfectly prepared to deal with any report that might be made against the Major 'upon its merits and apart from any previous transactions'.

(From this it may be assumed that Swindley's remark that 'he never forgave' was not an idle boast.)

By early October the Horse Guards had received the first three instalments of Sir Hugh's laboriously compiled *History of the Inniskilling Dragoons,* as well as the complaints regarding the criticisms made in Parliament and Press. On the 8th a reply was sent. It began:

'H.R.H. cordially approves of your offering of the fullest explanation if you consider that any part of the case has been mis-stated, that your motives have been misinterpreted, or your actions misunderstood; and he appreciates most fully the high military spirit in which, with every propriety, and with becoming frankness, you have approached a question *which, from the point of view in which you have regarded it, appears personal to yourself.'*

His speech in Parliament, was intended to dispel the popular impression that he could have summarily removed Lt-Colonel Crawley from his command in spite of the fact that his conduct had been approved by Sir Hugh Rose. To do this he had to explain how the executive authority in India was vested in the C.-in-C. there, and that only in exceptional cases, and even then in such a manner as to avoid compromising that authority, would he exercise his jurisdiction as final referee.

For this reason, while differing entirely from the views expressed in Sir Hugh's General Order, he had refrained from dismissing Colonel Crawley from his Regiment, although in his opinion that course was highly desirable. He had defended Sir Hugh's character and reputation because he believed his decision had resulted from an error of judgement 'to which we are all liable', and from a desire to do justice and uphold discipline.

He regretted that all the questions raised in England over the proceedings at Mhow had not been examined before they left India. For there would then have been no room for that wide difference of view which had occurred between the superior military authorities in the two countries.

On Lilley's character he refused to be drawn into discussion, but continued to state his opinion that the evidence on which Sir Hugh based his remarks appeared to be doubtful 'whilst the evidence of Lilley's character seemed unimpeachable'.

He went on: 'Moreover, H.R.H. has never ceased to entertain the grave objection which he conceived from the first to the introduction by your Excellency into remarks upon the trial of Paymaster Smales of reflections upon the character of a non-commissioned officer, not under trial, who, to say the least, had died under very lamentable circumstances.'

These then were the opposing views held by the two C.s-in-C. on the eve of the Court-Martial on Lt-Colonel Crawley.

19

The Charges are framed

Of all the problems thrust upon the Authorities by Crawley, the most exacting was what to charge him with and precisely how to word the charge. But it was entirely appropriate that the Judge Advocate-General, who had so unerringly laid his finger on the weakness in this pre-requisite to the Mhow Court-Martial as the prime reason why the proceedings had been vitiated, was now called upon to carry out this delicate task. And however alert he may have been to the danger of being hoist with his own petard, he was soon to receive a foretaste of the extreme difficulties in combining precept with practice when dealing with the military system of justice.

On 23 June he despatched to the Horse Guards a copy of two charges he had drawn up. But six days later he sent in another which contained 'a slight verbal alteration to meet the wishes of others'.

The reason for the amendment is not hard to find—the first referred specifically to Sir William Mansfield. It charged that after having received an order from Lieutenant-General Sir William Mansfield, KCB, to place R.S.M. Lilley under close arrest, Crawley caused the said order to be carried into effect with unnecessary and undue severity whereby R.S.M. Lilley and his wife were subjected to great and grievous hardships and sufferings.

The second version left out all reference to Sir William and substituted 'when R.S.M. Lilley was confined in close arrest'.

The second charge remained unaltered. It accused Crawley of

having stated in his address to the Court at Mhow that no one could have been more shocked than he was at hearing that Mrs Lilley had been incommoded by the precautions taken for her husband's safe custody and that, if it occurred, it was the fault of Lieutenant and Adjutant FitzSimon who should have taken care to see that the sentry was placed so as not to interfere with Mrs Lilley.

Whether Mr Headlam was to be more successful than his counterpart at Mhow had been in focusing the Court's attention on an issue on which the accused could be reasonably convicted remains to be seen. The 'slight verbal alteration' he had been persuaded to make certainly contrived to produce a feature he had disapproved of in the Mhow Trial—the introduction of a great deal of extraneous matter into the proceedings.

20

The witnesses are chosen

Although the trial was to take place in England, most of those who would be required to give evidence were still in India. Instructions on the important matter of the selection of witnesses were sent by the home authorities at the beginning of July 1863, together with the names of seven whose evidence was regarded as essential for the prosecution. They were Lt FitzSimon, Major Swindley, Q.M. Wooden, Surgeons Turnbull and Barnett, Captain Weir and Sergeant-Major Cotton. Of these FitzSimon, Swindley and Barnett were already in England.

In addition, plans of the rooms in which Lilley was confined were to be sent and 'someone' was to make enquiries on the spot as to whether anyone else could give material evidence. Crawley was to be allowed to call any witnesses he needed for his defence.

The people whose duty it was to carry out these, perhaps inevitably vague, instructions, were, of course, those who had most to gain from Crawley's aquittal. And in the light of past events as well as the natural desire for self-preservation, it would be surprising had they displayed absolute impartiality. However, almost all the information on this important subject has to be guessed at beneath the gloss of official correspondence.

On 25 July Sir William Mansfield informed Major-General Green at Mhow, of Crawley's impending trial and passed on the orders he had received from England. His letter contains one item of particular significance.

'It is probable,' he wrote, 'that it may occur to you that the evidence of Major Champion, Assistant Adjutant-General, may be useful at the approaching trial *on one side or the other*. If so, he might be most advantageously used with regard to the care of the models of the quarters, and providing their accuracy in Court from his personal acquaintance with the premises concerned. He might also be charged with such documentary evidence as may be deemed by you necessary to send to England.'

Now Major Champion had been Farrell's right-hand man during the whole affair under investigation and was personally involved in all that had taken place. His explanation of the models might be rather different from that to be expected from one of the seven key witnesses suggested by the Horse Guards.

On 15 August Crawley sent in a list of witnesses for his defence. In his short covering letter he said he believed some of those named were also required to give evidence for the prosecution, adding that the Major-General would 'probably deem it expedient to place them in that list'. This suggestion may have been prompted by his experience at the Mhow Court-Martial, when so many of his witnesses had afterwards so disconcertingly given evidence for Smales. In any case, the Indian Authorities must have recognized the advantages of having witnesses who intended to support the defence called in the first instance by the prosecution.

Crawley also wrote that his friends in England had informed him he might be arrested on arrival there and asked for the whole Regiment to be posted home on the grounds that he would require many more witnesses for his defence if he was to be tried in a civil court. He could not in those circumstances possibly select them, were the Regiment to remain in India. This request received some support from his superiors but was eventually turned down.

Two days later Major-General Green forwarded a list of prosecution witnesses to Poona, chosen for their ability to speak more positively as regards the Prosecution, though several of them *would be called for the Defence*. He also sent the numbers involved by telegram, to Sir Hugh Rose at Simla—5 Sergeants, 12 Corporals and 2 privates out of 7 Sergeants, 19 Corporals and 51 privates who had been investigated.

The shortness of the list evidently alarmed the C.-in-C. who immediately telegraphed Sir William Mansfield saying he could not possibly sanction keeping back any soldier who had stood guard over Lilley.

On 7 September Sir William supplied the Horse Guards with an amended roll of witnesses who were about to leave for England. It bore the names of 12 officers, 61 N.C.O.s and privates, one woman and 3 natives. They were expected to sail on the P. and O. Steamer *Salsette* on the 22nd, but the available accommodation proving insufficient, only half the party left on time, the remainder following on the next boat a week later.

The roll carried a footnote that one of the natives, the assistant to the sutler who supplied Lilley with goods including drinks, might not sail, and that although he was an important witness, the law of the country was powerless to make him embark. In the event, the man did not sail, though whether he actually refused or was 'persuaded' remains unknown. He was certainly an important witness.

Sir William also wrote that Major Champion would bring with him not only the plans and models of the buildings in which Lilley was confined but originals or copies of all the correspondence connected with the 6th Dragoons from their arrival at Mhow, and 'should his Royal Highness desire it, the entire subject, intimately connected as it is in all parts, may be put before him by an officer thoroughly competent from his position and knowledge to do so'. The role intended for the Major was now beginning to assume its true significance.

While the witnesses were being selected, a letter on the subject, 'From our own Correspondent', appeared in the *Deccan Herald*, which had throughout the affair been publishing information hostile to Crawley. It was now suggested that he was being allowed to take to England any witnesses he chose, whereas his opponents were not. It went on to allege that Mrs Cotton, the wife of the Sergeant-Major, whose testimony was said to be indispensable by Major Maude* who examined her, was not being allowed to sail. On 14 September Sir William Mansfield wrote to Mhow calling for a report on the letter so that if

---

* The deputy Judge Advocate-General indicted by Smales for helping Crawley to prepare his case at the Mhow Court-Martial.

inquiries were made he could give a full explanation *and refutation* [sic] of the allegations.

Major-General Green replied that 70 or 80 people had given statements which he and Maude had gone over together to select those witnesses able to 'speak most clearly and closely as to what the charges alleged'. The allegation regarding Mrs Cotton was most unfounded. Major Maude had questioned her because she was a friend of Mrs Lilley and had attended her in her illness. But he had not thought it necessary to send her home because her evidence was not of great importance. Green added that Mrs Cotton had, nevertheless, accompanied her husband to Bombay without orders and that the Sergeant-Major had written to his C.O. at Mhow for permission for her to continue the journey with him to England.

Colonel Prior, who had been left in charge of the Regiment, supported the request and Green (faced with a virtual fait accompli), was obviously accepting the situation with as good a grace as possible. He wrote, '. . . I, knowing that she had been a friend of the late Sergeant-Major and Mrs Lilley, and believing that she was with the former when he died, telegraphed to you recommending that Mrs Cotton be sent home.'

From the determined behaviour of Mrs Cotton and the evidence she would give, Green could hardly have been in doubt as to her being present when Lilley died, or as to her account of what happened.

Duval, the only one of the confined Sergeant-Majors still with the Regiment was one of several witnesses whose evidence would have been of particular interest. The circumstances of how his statement was taken are curious. According to Maude he was approached by the Adjutant (Lieutenant Devany) and asked if he had taken the testimony of Sergeant-Major Harvey (late Duval),* as he believed he could give evidence in favour of the prosecution. Maude replied he had not yet done so but if ordered he would. Devany then remarked that he hoped Harvey would not be required *as they wanted to make him R.S.M.*

Eventually, when Harvey was sent to make his statement he carried with him a letter from the Adjutant saying he did not

---

* 'Duval' was, presumably, an assumed name on enlistment and the S.M. had reverted to his real name of Harvey.

appear to know as much as was first supposed. After questioning him, Maude agreed that this was so. Harvey (late Duval) remained in India.

Maude's apparent ingenuousness also stretches credulity in the case of Pte Walsh, Crawley's ex-servant who, it may be recalled, had been threatening witnesses during the trial of Smales. He said he knew nothing of the man nor had his name been mentioned. Had it been, any unprejudiced person would know that Major-General Green would at once have ordered him to make his statement. Why anyone should suggest that Walsh, merely because he had once been Colonel Crawley's servant, ought to be ordered home as a witness for the prosecution or the defence, he was quite at a loss to understand.

21

The Press divides

In the autumn of 1863, several 'on the spot' accounts of conditions at Mhow appeared in the British Press tending generally
to play down the charges Crawley had to face.

On 4 November an anonymous letter sent at the end of August
from India appeared in *The Times*. The writer said he had visited
Mhow and spoken to people to check the facts about the imprisonment of Lilley. He concluded that the accusations against
Crawley were not borne out by the evidence and found it hard to
understand why the sophisticated members of society should
have accepted them so uncritically.

'The cry of "baking a sergeant to death" was naturally to
excite popular indignation. That it should be echoed, however,
and an ex parte and prima facie so improbable a story should have
been entertained and acted upon by educated gentlemen whose
public position seemed to demand a degree of judicial reserve, is
a circumstance less intelligible.'

A more detailed account was contained in an article in the
Illustrated London News. It was signed 'Hartley Hall' and there
were several illustrations by the author. His visit was made soon
after Crawley and the Trial witnesses had sailed for England
and he gives a cosy glimpse of Mhow Society in a state of armed
neutrality, politely leaving Crawley's name out of all conversation.

The men of the Regiment inclined warmly to their Commanding officer and hoped he would be cleared of the obloquy
attached to him. Lilley was a man without an enemy. When

arrested he had not been placed in the building usually reserved
for men under arrest but in a large room divided up expressly
for him, almost thirty by sixteen feet.

The N.C.O.s and men had raised an expensive but badly
designed tomb for Lilley as a token of their respect. It bore the
names of himself and his family. A story that when Crawley saw
the work in progress he ordered it to be stopped, was unlikely;
the stoppage was probably due to lack of money or shortage of
labour.

The article is interesting but the overall picture it presents is
too glossy. The scenery, climate and amenities are near to perfect. The officers ('Captain Sabretashe' and 'Major Shako') are
conventionally gallant and comic, and the men are conventionally
good-hearted and servile (as in a remark quoted on Lilley, 'He
was as good a fellow, Sir, as ever God put breath into').

The writer was either ignorant of Army procedure or deliberately misleading, otherwise he would have known that a
sergeant-major under arrest was not put in prison but confined to
his quarters. Nor was he curious enough to find the truth about
the delay over the monument to Lilley, the erection of which at
least signified a genuine emotion shared by the men of the Regiment and which could have produced a story for any journalist
worth his salt.

Everything suggests that Mr Hall was well taken care of and
well primed by the military authorities, that the officers of the
Regiment had no intention of expressing their views to a
stranger who intended to retail them to his readers in England,
and that the men had similar inhibitions or had been told to
mind their tongues.

Yet the piece was excellent propaganda for Crawley who was
now back in England, and was well timed to increase the swing
of influential opinion in his favour.

Weightier propaganda for his opponents appeared in a
twenty-five-page article by J.O. in the October issue of *Cornhill
Magazine*. Based on the documentary evidence available to the
writer, it was a masterpiece of advocacy and was to have a considerable influence on the forthcoming Trial.

The following extract suggests the likely effect on its readers
and also gives a piece of information which, though used to

good effect was to become a grenade to be tossed back by the enemy.

J.O. recounts the deaths of Lilley and his two children. He then introduces a letter written by Mrs Lilley the day before she died which, he says, could hardly have come from the wife of 'a dangerous and foul-mouthed conspirator'.

My Dear Brother and Sister,

　This is indeed a painful moment—a task I never expected to have to tell you. My beloved husband is no more. He died of apoplexy on the morning of 25 May. It was so sudden; he was tolerably well the day before. Dear sister, in mercy go to our father and mother, I cannot write to them. The blow will be too much for them. I am staying with Sergeant-Major Cotton and Mrs Cotton. I was to have gone into hospital but doctor says I shall not last long, so I don't think I shall be moved before anything happens. I cannot write any more; I cannot sit up. My best love. Your loving and affectionate sister,

Clarissa Lilley

　'A complaint has recently been made that the British public has reserved all its sympathy for the Lilley family, and has abstained from any expression of regret at the severe domestic affliction to which Colonel Crawley alluded when he complained to the court-martial at Mhow that "the upas-like shadow of Captain Smales had fallen on his threshold, and had converted his heretofore happy home into a scene of mourning and woe". It is perhaps as well, therefore, to explain that Colonel Crawley was bereaved of his mother-in-law whilst the Mhow court-martial was sitting, and that he thus informed Colonel Payn and his colleagues that the unlucky paymaster ought to be considered responsible for the melancholy event.'

　The mother-in-law gibe was well calculated to amuse the anti-Crawleyites. Yet it suggests that J.O., like most liberal and kindly men who commit themselves to a cause, was ceasing to allow his opponent any of the normal human emotions.

　A 'backlash' in the form of a reply by Thomas Hughes* appeared in *The Spectator* at the end of October.

　With J.O.'s general object, he wrote, every liberal must agree—that the proceedings of courts-martial were a great

---

* *The Spectator* is unable to identify the writer. He was probably the author of *Tom Brown's Schooldays*, who was a barrister. (Coincidentally, Bennitt went to Rugby!)

public scandal. But with his special object, that of drawing up an intelligible narrative of the events surrounding the Mhow Court-Martial, the case was very different.

Had the narrative been as fair as it was intelligible and able, it might have served a good purpose. But if J.O. had held a brief for the prosecution he could not have arranged his material in a more masterly manner with a view to obtaining a conviction. It was a speech of a leading counsel, delivered before the court was sitting; a sensation paper likely to deprive the accused of what chance he still had of a fair trial.

It was all the more unfair because the tribunal, as J.O. pointed out, was so weak. There would be no first-rate judge to keep the court to the points at issue, and the Court would be composed of men 'unused to sift facts, members of a profession notoriously sensitive to the applause and censures of the press.' Yet here was J.O., a trusted guide, saying, 'This man who is about to be tried is the guilty man.'

The writer then attempts to refute J.O.'s conclusions on the Mhow Trial. He quotes from the proceedings (mainly statements by Crawley) to show the bad state of the Regiment and goes on to contend that Crawley proved beyond doubt he was present at the muster parades. It was, he said, a soldier's rather than a civilian's question whether he was justified in signing himself present. But as a matter of common sense and common honesty he was undoubtedly present and entitled to sign.

He then says: 'I forebear from saying a word as to the arrest of the sergeants or the conduct of the trial. These will be the subjects of inquiry at the coming court-martial. But I must deny that their evidence which was actually taken on these points could have altered the verdict. . . .

'J.O. claims to be pleading the cause of the weak against the strong. It appears to me that "the weak" is the man the Horse Guards are prosecuting and against whom the press of the country are banded together. . . . Let us hold our judgements till we see what his case is as it will be established under the eyes of the nation at the forthcoming trial.'

To this a reply, written by J.O. from Naples, was published in the December *Cornhill*. The reader will judge for himself the relative merits of the cases presented by the two writers. But

it is worth noting one misapprehension of Thomas Hughes—
that military men were 'notoriously sensitive to the applause and
censures of the press'. On the evidence of the Mhow Court-
Martial at least, a good case could be made out for the opposite.

J.O. wrote: 'Mr Hughes, a most accomplished and honest
gentleman, and a very good friend of mine, has published in the
Spectator of 31 October, a criticism of my paper on the Mhow
Court-Martial. Mr Hughes in exposing what I am sure he
sincerely believes to be the unpardonable unfairness with which
my paper is written, speaks so handsomely of me, its writer, and
pays so many compliments to my pen at the expense of my con-
science, that it is with great reluctance that I dissect his article at
all. But I feel that I owe a duty to those whose cause I have under-
taken to advocate; and were I to kiss in silence the rod with
which Mr Hughes has so rashly smitten me, it might be supposed that I admitted the accuracy and justice of his criticism, in
which case others would suffer as well as myself.

'I think, therefore, that I had better proceed to show, as I very
easily can, that Mr Hughes does not understand the subject on
which he attempts to confute me; that he has not taken reasonable
pains to acquaint himself with the details, and that the arguments used tend rather to establish than to subvert the case as
stated by me.'

Mr Hughes, he said, had ridiculed as false his assertion that
the morale and discipline of the Regiment were good when
Crawley joined. But the Duke of Cambridge had stated in his
Memorandum that he had official proof it was in a satisfactory
state at that time. In addition, Crawley had himself admitted as
much in his closing speech at the Court-Martial.

As to Crawley's alleged presence at the Muster Parades, his
critic had plunged rather wildly and unintelligibly into the
evidence, only to arrive at the same conclusion as himself,
namely that the Colonel was not actually on the parade ground.
But, said J.O., '. . . he pleads on the Colonel's behalf that he was
in the neighbourhood and came on the ground soon after the
parade was over. . . .'

The fact was, Smales had not charged Crawley with neglecting
his regimental duties. He merely said that whenever he had
deviated in the smallest from his, Crawley had brought formal

charges against him. He went on to allege that Crawley had also deviated by absenting himself from muster parades. Mr Hughes had carefully studied the evidence and decided the allegation was true. Where then was the 'falsehood and malice' with which the Paymaster was charged?

J.O. concluded: 'I have only one more observation to make. Mr Hughes accuses me of having improperly and unfairly prejudged the case about to be tried at Aldershott; the subject of which is to be, according to Mr Hughes, "the arrests of the sergeants, and the conduct of the court-martial at Mhow". Now there can be no denial that both these points have been effectually prejudged by far higher authorities than myself—viz., H.R.H. the C.-in-C. and the Judge Advocate-General. But it will scarcely be believed that Mr Hughes has ventured upon his public criticism of my paper, and his public condemnation of myself, without taking the trouble to inform himself as to what the charges really are upon which Colonel Crawley is about to be tried. If he will inquire, he will find that they have nothing whatever to do either with the arrest of the sergeant-majors, or the conduct of the late trial.

'The approaching inquiry is limited, first as to whether Colonel Crawley carried out the orders of his superior officers in the arrest of Sergeant-Major Lilley with unnecessary cruelty; and secondly, as to whether he falsely stated that the alleged cruelty with which Lilley and his wife were treated, was to be attributed to the misconduct of his Adjutant, not to any orders given by himself.

'I invite Mr Hughes to cite a single sentence in my paper in the *Cornhill*, which prejudges either of these two points.'

Judged according to the skill in marshalling evidence from the proceedings of the Mhow Court-Martial and its sequel, there could be little doubt as to who had come off best in this verbal contest. But the moulding of the blunt instrument of public opinion depends on far more subtle influences than mere logical argument. This was recognized by a commentator in the *Illustrated London News* on 12 December when he wrote, '. . . and however dextrously J.O. may avail himself of weak points in Mr Hughes' reply to his paper on the Crawley case, he cannot deny that it was calculated to embitter public feeling respecting the Colonel at a time when it was most important that it should be, as far as possible, unbiassed.'

With motives that were admirable and a case almost perfect, perhaps too perfect, for once J.O. was at fault because his timing was wrong.

22

The Case for the Prosecution

The venue decided on for the Trial was the Club-house at Aldershot, on the right of the Farnborough Road, opposite the South Camp. The building, agreed to be admirably suited for the purpose, was described by the *Illustrated London News* as 'of iron construction with a centre and two wings, not unlike a mission church'.

The main room was converted into a court-room, two-thirds of its area (70 by 39 feet) being allotted to the court which was divided from the public section by a barrier. The heating and ventilation during the trial was said to compare very favourably with that in other courts (and was far better than in the House of Commons). On 3 December, however, gale force winds 'rarely experienced even in Aldershott' so pounded and shook the iron structure as to make it almost impossible at times to hear what was being said.

The short winter days also exposed the inadequacies of the lighting system which consisted of clerestory windows and three large glass chandeliers. At the end of one day a witness was about to explain certain entries in a book when the President told him he would have to wait until the next morning as it had become so dark that nothing in the book could be seen.

On the morning of 17 November, the Court, composed of Lieutenant-General Wetheral, KCB (the President), four major-generals, four colonels and six lieutenant-colonels, duly assembled.

To avoid the unpleasantness of moving to and from the building, apartments had been fitted up for the Prisoner in the Club-house.* So, with a minimum of fuss and punctually at 11 am, he entered the Court-room, bowed to the President and was con-ducted to his seat.

At his table were Mr Vernon Harcourt[13] (as yet a compara-tively unknown young barrister), Mr Waller of the Chancery Bar, Mr Cocker his solicitor and Lieutenant-Colonel Henry Crawley, his brother.

At a table on the other side of the President sat the Prosecutor, Colonel Sir Alfred Horsford,[14] Mr Denison of the Judge Advocate-General's Department and their assistants.

After hearing the Charges, Crawley raised an objection. They were, he claimed, so limited and vague that he had been seriously hampered in preparing his case. He wished to lay before the Court a series of letters concerned with his requests to the Horse Guards to have the Charges enlarged to enable him to give full explanations of everything that had happened and so prove to his countrymen how badly he had been misrepresented. But his requests had been turned down and he 'bowed with humility to his Royal Highness' decision.'†

He then formally pleaded 'not guilty' to the Charges.

Horsford, in a short opening address said the Prosecution intended to prove that Lilley and his wife had been subjected to 'great and grievous hardships and sufferings' and that these were caused by the severity of the conditions of the arrest for which the Prisoner was responsible.

The proofs they would bring forward would include plans and models of the buildings in which Lilley was confined, the posi-tion of the sentry, the constitutional habits of the Sergeant-Major, his wife's illness and the medical evidence on the results of the confinement.

They did not impute blame to the Prisoner for the length of the arrest as this had been sanctioned by his superiors. But its

---

* See Note 12 for the consideration given to a private when being tried and con-victed.

† Crawley had not been alone in wanting the charges altered. Thorndike and Smith for the Lilleys had submitted two additional ones to the Horse Guards. The first, that Crawley be tried for the illegal arrest of the Sergeant-Majors; the second for prohibiting a female nurse from attending Mrs Lilley. Needless to say they too had no success.

conditions should have been 'as little grievous as possible' from its commencement. And as time went on the Prisoner should have taken steps to find out how long it was intended to keep the Sergeant-Majors in close arrest.

The medical evidence would show that excessive drinking had in no way accelerated Lilley's death. There was no medical reason why he would not have been in good health at the time he died had it not been for his confinement. Clearly, the hardships and sufferings that so contributed to the death of a very strong man must have been great and grievous.

Up to 7 May, when Lilley gave his evidence at the Mhow Court-Martial, the orders regarding the placing of the sentry were 'so positive and precise' that it was unlikely they would not have been put into effect.

There would be proof that these orders were given by the Prisoner in the Orderly Room to Lieutenant FitzSimon in the presence of several witnesses a few days before 7 May and that FitzSimon and Sergeant-Major Cotton raised objections to them.

If this was proved to the satisfaction of the Court, it would be for them to consider whether Lieutenant FitzSimon could have been to blame for what took place.

FitzSimon was then called. He gave his evidence as follows:

'On the 26 April, 1862, I was sent for to Lieutenant-Colonel Crawley's house. Colonel Crawley asked me if I knew whether there was a conspiracy going on against him. I said I was not aware of it. He said I did not do my duty as Adjutant; I should know everything. He told me that R.S.M. Lilley, T.S.M. Wakefield and T.S.M. Duval were in conspiracy against him. He ordered me to place them under arrest, which I did. I asked him if sentries were to be placed over them. He told me he would let me know later, as he first wished to speak to the general about it. Subsequently he gave me a written order to place sentries over them.'

The order read:

'R.S.M. Lilley, T.S.M. Duval and T.S.M. Wakefield to be placed in close arrest, with a sentry placed over the quarters occupied by each of them, with orders to allow no person whatever to hold communication, verbal or written with the prisoners until further orders.'

FitzSimon went on, 'A few days subsequently Colonel Crawley said to me in the Orderly Room, "Persons have had intercourse with the prisoners", and he asked me if I knew what close arrest meant; he said that close arrest meant a sentry was not to lose sight of his prisoner day or night, and he gave me orders that the sentry should be placed inside.

'Acting R.S.M. Cotton made a remark to Colonel Crawley that Lilley was a married man. Colonel Crawley answered to the following effect—officer or soldier, married or single, I do not care a damn (I believe that was the word he used), the duty should be done. I also remarked, to the best of my recollection, that Mrs Lilley was sick and that her husband was obliged to rub liniment on her chest every day.'

FitzSimon then said that on 4 or 5 May he was 'invalided' and Cornet Snell became acting Adjutant. He resumed duties on 16 May but was suspended on the 22nd.

Questioned on his attempts to prove that his eyesight was not defective, his account of what happened after he had sent in his letter denying Crawley's allegation of short-sightedness agreed substantially with that given by J.O. in his letter to *The Times*.

The next witness to the conversation in the Orderly Room, Sergeant-Major Cotton, said that when he reminded Crawley that Lilley was a married man, Crawley asked him how he dared to disobey his, or the Major-General's orders and told him to go and see that the sentries were properly posted inside the quarters.

Six more witnesses swore to being present at the all-important conversation and gave accounts that varied only slightly in detail. They were, Major Swindley, Captain Weir, Q.M. Wooden, the Orderly Room clerk and two sergeants.

That Crawley intended the order for close arrest to be carried out to the letter was shown by what happened to two soldiers who failed to do so.

Pte Blake told the Court that as corporal of the guard he posted the sentries outside the door and was court-martialled for neglect of duty. From the proceedings of the trial, put in evidence by the Prosecution, it appeared that he had received verbal orders from a sergeant named Mills (of whom more later) to place the sentries inside the quarters of the Sergeant-Majors. But just before midnight on 1 May they were all found outside.

Blake gave the excuse that there was no light inside, which suggests he may have been using his initiative—a dangerous course in the circumstances.

He was sentenced to 42 days' imprisonment but with a recommendation to mercy. This was acted upon by Crawley, who reprieved him on the grounds that he had acted in ignorance rather than with criminal intent and he was allowed to rejoin the Regiment as a private.

The second soldier, Pte Little, was arrested on 8 May for allowing Mrs Gibson (the wife of Sgt Gibson) to talk with Mrs Lilley on the verandah of the first bungalow. He was not punished however, as Major Swindley 'let him off' (much to Crawley's annoyance).

Further evidence of Crawley's responsibility for the conditions of Lilley's arrest was given by Cornet Snell who became Acting Adjutant when FitzSimon went sick. He said that after Lilley had given evidence in the Court on 7 May he was sent for by Crawley and asked if it was true the sentry was within two feet of Mrs Lilley's bed. He replied that the sentry was inside the bungalow next to Mrs Lilley's bedroom.

Asked who gave the order, he said 'I understand it was yours, Sir'.

Crawley said, 'I ordered close arrest; why are not my orders carried out?'

Snell then left the Mess, went to Lilley's bungalow and placed the sentry 'in the lobby or small room, next to the dining room with the order that he was in no way to interfere with Mrs Lilley, but not to lose sight of R.S.M. Lilley except when he went into his wife's bedroom.'

After the Court-Martial had adjourned he went to Crawley's bungalow with regimental papers and handed him a written copy of the order he had given the sentry. Without reading it Crawley passed it back saying, 'Do you think, Mr Snell, I am a lance-corporal to write out orders for a guard?'

Snell explained he had taken the order to Crawley for his approval because he had only been Adjutant for a few days and did not like the responsibility for changing an order that was in force when he took over the duty.

The next day when he went with the regimental morning

II

papers Crawley told him that Lieutenant Davies would take over from him as Acting Adjutant and that he was to return to his normal duties.

Of the exact orders given to the sentries, two corporals gave versions varying only slightly in phraseology. They said he was not to lose sight of the prisoner or allow communication to be held with him. He was to allow no one into the quarters except the medical officer and the native servant. The latter was to be searched coming in and going out to ensure that he had no written documents hidden about him. Any such found were to be taken from him and given to the sergeant of the guard who was to give them to the Adjutant for the C.O.'s information.

At a more personal level, Pte Atkins, an old soldier of twenty years' service, who had known Lilley since he first joined, gave the following account of an incident when he was on sentry duty.

'I was on sentry between 8 and 10 o'clock; on the first day I was over, him and Mrs Lilley was dusting the things, the ornaments in the large room. He told her not to be knocking about like that, that she would do herself harm. In a few minutes she fainted on the sofa. The Sergeant-Major went and got her some soda water. She drank a little of it and threw it up. Then he gave her brandy three times; she threw it up twice; the last time it remained on her stomach.'

Asked whether Mrs Lilley knew of his being on sentry and whether she did anything in consequence, he said, 'I don't know whether she did on that occasion, but she did once before because she asked the Sgt-Major who it was. He said it was Atkins. She said she did not mind about me; she would come outside as I was a married man.'

A private named Gaffney also described a personal experience. On one occasion he had been on duty about half an hour when: '. . . Sergeant-Major Lilley came to me and asked me to do him a kindness, to go outside the bungalow while his wife undressed out of the room; he wanted to rub some liniment into her breast. I did so. When his wife was dressed again I came back to my post.'

Horsford then asked Gaffney if he was a prisoner. He replied,

'Yes. For drunkenness . . . I have been eight and a half years in the regiment and all of the times (in trouble) for being drunk.'

Instead of cross-examining this witness, Crawley protested that he was not on the list of Prosecution witnesses given to him.

Horsford retorted that the prisoner had named all the soldiers who came from India as his witnesses and had the same access as the Crown to Gaffney who had not been included in the list because he was under arrest and he (Horsford) had not known of him until the last moment. The list had been given to the Prisoner in the usual way, but if another witness appeared there was no obligation to tell him of it, nor was this ever done.

The Gaffney incident is remarkable as perhaps the only occasion when the Prosecution managed to steal a march on their opponents.

As it was hardly disputed that the sentries were ordered to keep Lilley continually in sight, it seems strange that a good deal of questioning was done as to whether they could actually see him through a screen known as a 'chick' from where they were posted.

Major Champion had the foresight to bring one of these 'chicks' from India and produced it for the inspection of the Court. It was constructed of split bamboos bound together with twine and lined with calico to obstruct the vision; though whether this was typical of all 'chicks' may perhaps be doubted.

Some sentries, perhaps unwilling to admit the fact, said they were not able to see through into Mrs Lilley's room. But a few admitted being able to see the bed if they went close to the 'chicky'.

Of Lilley's removal from the first bungalow, a different story to the official one (that Lilley was given the best quarters available when the building had to be pulled down to make way for new barracks) was told by Wooden, who, as Quartermaster, was of course involved in the transaction.

He said that both Lilley and the Quartermaster Sergeant had to leave their quarters because of the acute shortage of suitable accommodation. But a staff-sergeant's quarters had been made available to the Regiment about a quarter of a mile away and he had suggested to Crawley that as Lilley was not required for duty this might be allocated to him.

Asked if he could be placed anywhere else, Wooden replied that there was only the men's married quarters.

Crawley said that neither would do; the one was too far, the other too open and that Lilley would be up to some more tricks.

He told Wooden to move the Sergeant-Major of H Troop out of his quarters and put Lilley there.

Wooden remarked that the quarters were small but was told that the quarters good for one sergeant were good for another.

When he tried again with, 'But the Troop S.M. is a single man. . .', Crawley lost patience and ended the matter by saying 'I don't want any more buts; do as I tell you.'

(Crawley had one very obvious reason for refusing to let Lilley occupy the Staff Sergeant's quarters—they were not only outside the barracks but situated just beyond the house where Smales was living.)

A first-hand account of Lilley's sudden death was given by Mrs Cotton, the lady who had taken matters into her own hands by accompanying her husband from Mhow to Bombay without the consent of the military authorities.

She said: 'On the evening of 24 May, 1862, about 10 minutes past 10 o'clock, I was called on by Mrs Lilley to come into her quarters, as her husband was very ill and herself not able to move in her bed. I did not go when she first called, but about 10 minutes after she called again. I then went out and found Sergeant-Major Lilley very ill. He asked me to send for the doctor, which I did. I stayed with him until about 10 minutes to one. I then returned to my own quarters and remained about a quarter of an hour when Mrs Lilley called to me again and said her husband was dying. I immediately went in and found him still worse. He looked at me very hard. Mrs Lilley said, "Do you know who that is?"

'He says, "Yes, it is Mrs Cotton."

'He then took my hand and says, "Mrs Cotton, I am dying."

'I said, "Don't say so Lilley, you'll soon be better."

'He said, "No, my child, I never shall."

'He took my hand and said, "Goodbye, May God bless you."

'I then assisted Mrs Lilley from the bed where her husband was dying and laid her on a sofa till all was over. I then had her removed into my own quarters, where she remained until about a fortnight before her death.'

It need hardly be said that Mrs Cotton's evidence had no appreciable effect on that exclusively male assembly in mid-Victorian England or on the course of the Trial.[15]

Dr Barnett, the first of the medical officers to give evidence, said he had been attending Mrs Lilley for six months before the arrest of her husband. She had consumption and by that time was in a declining state.

In the first bungalow her bedroom was Room 3, the most convenient for her as it was contiguous to the room used as a water closet. Her bed's head was a foot or so from the door of Room 2 where the sentry stood and he could see into the room though not distinctly without moving the chick over the doorway. When he attended her he could see the sentry's head clearly over the chick. He considered the position of the sentry during the whole of the R.S.M.'s arrest must have been an annoyance to him and his wife and the alteration of his position after 7 May had not made the annoyance less. Seeing the sentry over the Sergeant-Major whom he had known so long and over his wife who was sick, made a very painful impression on him at the time.

He said he always spoke in an undertone because normal conversation could be overheard in the next room.

Mrs Lilley was often confined to her bed for whole days, but was sometimes much better and got up.

He never saw any appearance of drinking on Lilley either before or after his arrest. The first he heard of it was at 5 o'clock on 9 June, 1862, when he was riding to morning parade with Crawley who asked him if he had any idea how much Lilley drank during his confinement. He said he had not. Crawley then told him it was '23 bottles of brandy, 12 pints of ale, a bottle of port wine and some gin'.

He waited for Crawley after the parade to find out if he wished them (i.e. Turnbull and himself) to make an additional report on Lilley's case regarding the drink. This was eventually written by him but signed by his superior.

Turnbull, who had not attended the Lilleys in this period, told how, four days after the arrest, he recommended the

Sergeant-Majors should be allowed exercise. Then on 18 May, after Barnett had reported a request from the prisoners to be allowed to sit on their verandahs because their rooms were very hot and close, he visited Lilley to confirm this and went to put the request to Crawley. Before giving his consent, Crawley said harshly, 'Why they have morning and evening exercise. What more do they require?'

In view of the sudden death of Lilley he thought a coroner's inquest should have been held. He applied at the time to Lieutenant Davies, the Acting Adjutant, to ask Crawley if he wanted one. But Davies came down to the 'dead house' to tell him that Crawley did not think an inquest was necessary.

He gave an account of how the Additional Report came to be made and admitted, as Barnett had done, that his opinion of the cause of death was 'to a certain extent modified by it'.

23

Defence tactics

Crawley had routed his opponents at Mhow by exploiting the 'system' and from the weight of evidence mounted against him he would now have to repeat the performance. But before describing the main lines of his strategy, it is as well to reveal something of the opposition he had to face.

When Colonel Horsford was appointed Prosecutor, none but the most closely initiated would have given a second thought to the matter. Yet, in fact, he had been placed in a situation of almost impossible difficulty. For with a legal knowledge hardly greater than that of any other experienced regimental officer he was being required to act as leading counsel at what was virtually a State Trial.

Not only would he have to present a case of enormous complexity involving masses of conflicting documents, but to sort out the evidences of almost a hundred witnesses from whom no depositions had been taken and who were being sent from India on the supposition that they could 'contribute something' to the Trial.

Ranged against him was an eminent barrister assisted by another, a solicitor, and the Prisoner himself who knew far more about the case than anyone. Yet, by court-martial rules he was allowed no solicitor to sift the evidence nor was any professional lawyer deputed to help him.

In this predicament, Horsford had gone for advice to the only source open to him, Mr Denison from the Judge Advocate-General's Department, who had been appointed to advise either

167

side on legal matters and to decide any points of law as they arose during the trial. Denison, a qualified barrister, at first gave general and then particular advice, but before long found himself so involved in this part of his duties that he wrote to Headlam, his Chief, to say he could no longer carry out the part of legal adviser to the Court.

Headlam, now convinced that unless Horsford was given proper professional assistance, the Trial would be reduced to a farce, approached the War Office and the Horse Guards with a request that both Denison and a solicitor should be appointed to help to prepare and handle the case for the Crown. He argued strongly that only in this way could justice be served. But whatever may have been the personal views of the heads of these Institutions, he came up against the inevitable wall of tradition. There were, he was told, precedents for his Department taking part in prosecutions at courts-martial but none for the use of a civilian solicitor. To bring in such an outsider at a time like this would cause very strong resentment in the Army and rouse such bitter controversy as to bring the Trial into disrepute from the start.

Against his better judgement, Headlam had to agree. He released Denison to give Horsford all possible help and appointed as Judge Advocate in his place, Colonel Pipon, an 'expert' who had written a book on the subject. In this choice too, Headlam may have acted from necessity. He could hardly have failed to foresee that with learned counsel on either side, one or both would use all the legal niceties and tricks of their profession to gain the advantage. Without the knowledge, experience and authority of a judge, Pipon would soon be out of his depth.

The first day of the Trial had been mainly taken up with the evidences of four witnesses who appeared on the Prosecution List. However, two of these, Major Champion, so shrewdly 'planted' by the Indian Authorities as the custodian of their documents, and a sergeant named Mills, were in fact important witnesses for the Defence. The use made of them to frustrate the Prosecution in presenting its case illustrates the tactics employed by Crawley and his advisers.

Major Champion made his first brief appearance to certify the plans of the quarters in which Lilley had been confined. But instead of cross-examining him immediately in the normal way, Crawley asked permission to delay this until later.

Horsford, obligingly, raised no objection, merely expressing the hope that such a departure from normal procedure would not be repeated. The request was granted. Then, as each succeeding witness concluded his evidence, Crawley applied for and was allowed the same concession.

By this device the Defence gained considerable advantages. The legal experts were given time to sift the evidence and indicate the best lines of questioning to their client. And, from the Prosecution's viewpoint, when the witness reappeared, often a day later, the impact of much he had said was inevitably weakened.

When Champion was recalled for 'cross-examination' the following day, he gave glowing descriptions of the quarters occupied by the Lilleys and, after legal argument, was allowed to introduce a number of letters including one from Sir William Mansfield sanctioning the arrests of the Sergeant-Majors.

The President commented that this appeared to have nothing to do with the Charges. But Crawley argued strongly that he had to establish the reasons for the arrests in order to explain the severity with which they had been carried out.

Horsford at first insisted that the Charges had been carefully framed to exclude the grounds of the arrests and that if they were gone into there would be no end to the trial. But again, with remarkable generosity, he ended by saying he would not object to the introduction of any documents which bore, *however indirectly*, on the issues raised. It was a gesture he had ample time to regret as the Trial progressed.

The President agreed that the documents were irrelevant. The question was whether the Prisoner had employed undue severity in carrying out the arrest; the Court would assume Lilley 'to be a man of the worst character'. This last statement was accepted by Horsford (though certainly not by Crawley) at face value for most of the trial, as meaning that Lilley's character was irrelevant to the Charges.

When Horsford rose to question Champion on the documents just introduced, he received a further indication of how seriously he was to be hampered by the confusion of witnesses. For

Crawley at once objected, claiming that Champion had been called by the Prosecution; the Prosecutor could not possibly be allowed to cross-examine his own witness.

The President rightly pointed out that the Major had just been questioned about documents which the Prisoner had put in for his defence, thus making Champion his own witness. But Crawley disagreed, setting in motion another of the long debates intended to confuse the issues.

Its conclusion, too, showed a familiar success for the Defence, with Horsford stoutly maintaining his *right* to question the witness but announcing that he did not intend to avail himself of it on this occasion.

What Horsford's failure to cross-examine with vigour must have cost his case is incalculable though it is suggested by a rare instance in which he did score a partial success. The witness was again Champion, who made several appearances for the Defence.

He had said previously, in reply to Crawley's questioning, that Lilley's quarters consisted of two large rooms as well as an inner and outer verandah, and that they were now occupied by a quartermaster-sergeant and his wife. Pressed by Horsford to repeat this description, Champion now altered it to:

'I am glad to have the opportunity of explaining that in addition to the quarters formerly occupied by Sergeant-Major Lilley, another room on the North side of similar dimensions to that in which he lived was in the possession of the quarter-master-sergeant.'

This retraction of his previous evidence removed all doubt that Lilley had been confined in a single room divided into living and sleeping quarters by a curtain, and that the sentry occupied the only other compartment available. Yet, having established this important point, Horsford did nothing to press the advantage by destroying the credibility of the witness. It was an opportunity that Crawley would never have missed.

The confusion of witnesses played into the hands of the Defence in yet another way. For though Horsford knew which officers would testify for him, he was far less certain when it came to the soldiers. This was painfully brought home to him when he asked Sergeant Mills, the second 'planted' witness, to describe Lilley's habits and character.

Mills said:

'He consumed a great deal of liquor; he drank very heavy and could carry a great deal. I should describe him as a coarse, ignorant man.'

In spite of this evident hostility, Horsford went on to ask (probably to show that Lilley's confinement had affected his health) whether he was a man 'of active stirring habits'.

Mills replied that '. . . he rode about a great deal on a pony, but he could not walk for he had bad feet'.

It must be assumed that Horsford was taken aback by these answers and that he had either not questioned Mills beforehand or had been led to expect quite different ones.

At all events, from then on, he avoided questioning many of the other ranks, passing them over immediately to Crawley for 'cross-examination'. Then, after several days of the farce, he admitted defeat by asking if the Prisoner could possibly cross-examine the remaining sergeants and sentries at once, as they were being called by the Prosecution 'with scarcely any other view than to be examined by the Court or the Prisoner'.

24

Cross-examination of the officers

Crawley had taken up all the second day 'cross-examining' Champion on the documents put in for his defence. He occupied the next three in subjecting FitzSimon, the Prosecution's key witness to cross-examination of a more orthodox kind, in the course of which his ex-Adjutant was manoeuvred into one damaging admission after another.

Asked if he had read reports in the papers that the second quarters occupied by Lilley was a bomb-proof oven, unfit for human occupation, FitzSimon said he had. But he conceded they were not true and that he had taken no steps 'to correct an impression so injurious to his Commanding Officer'.

He agreed that an adjutant was his C.O.'s right-hand man; yet he had partaken of Mr Smales' hospitality when that officer was on a charge of insubordination to Colonel Crawley. For this and his incompetence he had been dismissed from his adjutancy by Sir William Mansfield and censured by Sir Hugh Rose.

Asked whether he remembered being censured by the Court at Mhow for the evidence he gave, he said he had read some of the proceedings but repeatedly denied having read that particular part.

Crawley produced the Blue Book, had the relevant passage read out and demanded to know how he could possibly recall all the details of a conversation in the Orderly Room yet remember nothing of a censure on his conduct by a Court-Martial.

FitzSimon explained that when he read Sir Hugh Rose's

censure he intended to write a letter of explanation and made enquiries to confirm who was present in the Orderly Room.

'Then,' said Crawley, 'your statements as to what was said depended on information from others, not from your own knowledge?'

On the posting of the sentries, FitzSimon's replies were vague and equivocal. He had no distinct recollection of when he had visited them or where they were placed. He never went into the quarters to inspect the sentry because he knew Mrs Lilley was sick and did not want to intrude upon her. Instead he spoke to the sentry at the back door.

Crawley seized upon this to ask why, knowing Mrs Lilley was sick, he did not go into the house to post the sentry where she would not be inconvenienced instead of leaving where to stand to the sentry's own discretion.

FitzSimon replied that he considered the presence of an officer would cause Mrs Lilley more embarrassment than that of a sentry. He did not think it was left to the sentry's discretion as the orders made it essential he should follow the R.S.M. from one room to another.

It was a pointer to the way the Court was turning that it here intervened to ask whether he gave the discretionary powers to the sentry 'out of pure consideration for Mrs Lilley?' To which FitzSimon could give no satisfactory answer except what the Court had tacitly rejected.

Later he was asked whether he had consulted anyone before writing his letter of remonstrance to Sir Hugh Rose. He said he had asked Q.M. Wooden to assist him 'as he could write a letter well'.

Having elicited this further piece of evidence of the plot against his authority, Crawley now produced and read out a letter he had himself written at the time to refute the complaints of his ex-Adjutant. This, of course, had been accepted by the chain of command in India, whose opinion was bound to weigh heavily with his judges.

He turned last to the way in which FitzSimon's letter had come to the notice of the Horse Guards. Was it not unusual, he asked, for one officer to renew a charge against another after he had withdrawn it?

FitzSimon replied that it might be unusual but he had only withdrawn his letter after the Major-General pronounced it insubordinate and told him it would produce his professional ruin. He did not renew his charge until the Horse Guards called for his letter.

He believed Smales did forward his letter though he had no idea how it came into his possession. He did not send it to him and had never discussed it with him.

It is not hard to guess the interpretation the Court would place on these statements.

Major Swindley, the next officer to be cross-examined, was of a very different calibre from the nervous and ineffectual FitzSimon. Yet in the Court-room, Crawley's ability to handle him was as marked as had been his failure to do so in the Regiment.

With a few shrewdly directed questions he drew attention to the fact that the Court at Mhow had commented on Swindley's animus and went on to press him to say whether he had not expressed his hostility at other times.

'Did you not tell me on one occasion that you never forgave?'

Swindley appealed to the Court for protection as the question had nothing to do with his evidence-in-chief and was put only to 'damage his prospects'.

The Court deliberated. It then pronounced the question in order as it was intended to prove animus, adding with unconscious irony, that he *need* not answer if he felt he would incriminate himself.

Swindley admitted having made the statement.

He was then asked tauntingly:

'I may assume, I suppose, Major Swindley, that as you were at so much pains to alter the first account you gave of the conversation in the Orderly Room, that you are prepared to swear to the words, "I don't care a *damn*"? '*

* After giving his evidence, Swindley had returned later to insert the expletive into his account of the Orderly Room conversation. Why, is hard to say. Even he would hardly have so deliberately prejudiced his case by displaying such obvious animus. It was said later that he came back 'not of his own accord, but was called upon in terms which no officer or gentleman could refuse to obey, to go back and complete his evidence.' (Hansard, 15 March, 1864.)

Swindley said he was.

Asked whether he had been communicating with Mr Smales, he declined to answer.

'Is that on the grounds that the answer may incriminate yourself?' said Crawley.

'No, certainly not,' Swindley retorted, 'I do not consider it a proper question.'

Here it was indicative of the effect Swindley's uncompromising attitude was having on the Court that the President now told him he was *bound* to answer unless he would incriminate himself. But Crawley, having made his point, said he would not press the question.

Although Cotton's evidence on the sentries and the Orderly Room conversation was equally as important as that of the officers, Crawley's manner towards him was markedly different, as indeed it was to all the other ranks he interrogated. As on previous occasions he was intent on demonstrating how loyal and disciplined his men had remained in spite of the disloyalty and insubordination of his officers.

The means by which he now set out to achieve this was simple but effective.

The questions he put to the men were easily understood and required only short, straight-forward answers, whereas, in the words of *The Times*, 'the skill of the advocate was well nigh exhausted in the questions put to the officers'. As a result, the men gave prompt intelligible answers and appeared as honest witnesses, while most of the officers were perforce hesitant in their replies and gave the impression of shiftiness and equivocation.

In Cotton's case, Crawley may well have had to exercise considerable restraint, for the replies he received appear, in print at least, as forthright as any. But he carefully refrained from calling into question Cotton's motives, denigrating him or holding him up to ridicule.

For Captain Weir there was no such immunity. The details of his insubordinate letter on Brigadier Hobson's Inspection, the censure by the Mhow Court-Martial for his animus to Crawley and his alleged opposition to changes in the running of the Regiment, were all paraded before the Court in spite of a protest by Horsford that some limit should be set on such irrelevant matter.

In reply to the usual question he denied having had any communication with Smales and said they had only met casually in the Court when he was coming in to see someone else.

It seems obvious that Weir was trying to avoid any further involvement in the affair, and not surprisingly. Now near the end of his long service he had shown signs over the past year of a realization that bucking authority was both dangerous and unprofitable for one in his position.

Q.M. Wooden enjoyed a considerable literary reputation in the Regiment. He had undoubtedly written the letter of remonstrance sent in by Weir, had helped FitzSimon with his and had been used by Crawley himself to write reports on Smales. Inevitably then, Crawley, who regularly attacked his opponents' talents and virtues (Lilley's sobriety, FitzSimon's eyesight) now went for Wooden's prowess as a correspondent.

'Mr FitzSimon has vouched you as the best letter writer in the 6th Dragoons. Pray is that one of your letters?'

Wooden was shown a letter written by him in which he stated his impression of the remarks made by Crawley to the Regiment after Sir Hugh Rose's General Order on the Mhow Court-Martial had been read out.* (This too had been written on Crawley's instructions and shows a good deal of astuteness on Wooden's part in not actually criticizing his C.O. yet suggesting the speech was rather uncalled for.)

Cross-examined on the letter he said it had been composed partly from notes he had taken on events as they occurred because he felt sure he would sooner or later be called upon to certify them. Asked why he made the notes, he said they were for his own protection and to make his answers as truthful as possible as he was frequently called upon, sometimes at a moment's notice, to write statements on what had taken place months before. But when he read in an account of a recent court-martial that it was wrong for an officer to keep a diary he decided to destroy his notes and trust to his memory.†

Dr Barnett, when cross-examined, said he had been on

* See note 7.
† The Bentinck-Robertson Case. See Note 8.

friendly terms with Crawley during Lilley's confinement and was at his house day and night. (He was attending Mrs Taylor up to her death at the end of May.) A mutual silence had been observed on everything relating to the Court-Martial because he believed it would have been impossible to carry out his duties properly had he become a partisan of either Crawley or Smales.

From this Crawley could hardly have tried to accuse him of 'animus'. Instead he concentrated his attack on the events of the few hours when, torn as to loyalties and uncertain of his facts, Barnett produced his Addendum to the Medical Report.

Asked how he came to write that Lilley had drunk considerable quantities of brandy daily, Barnett said he had gone to ask Mrs Lilley if it was true. She told him it was.

The Prosecutor interjected that while he did not object to this kind of questioning, it involved the worst kind of hearsay evidence—conversation with a dead person. He would accept it only if all Mrs Lilley's statements on the arrest were admitted.

Crawley persisted that the question was necessary to refute the accusation that he was responsible for Lilley's death, but said he would substitute another.

He then asked: 'Of whom did you make your enquiries?'

Barnett said: 'Of the late Mrs Lilley.'

Horsford immediately protested that this was precisely the same as before. But he was over ruled and the reply was recorded.

Before he was allowed to stand down Barnett continued to be pressed with long involved questions designed to prove his negligence in his treatment of Lilley.*

Turnbull, under cross-examination, managed to insinuate that improper motives had driven Crawley to arrest the Sergeant-Majors. He said he had not recommended exercise for them at first because of 'the unusual and mysterious nature' of their arrest. He tried but was unable to find out why they were confined and assumed they were in close arrest because Colonel Crawley wished them to be so.

But as with all the important Prosecution witnesses Crawley

---

* See Note 16.

Though by modern medical practice his treatment must have hastened rather than retarded the Lilleys' deaths, it seems to have been open only to minor criticisms by that of his time.

found a point at which to hammer until a crack appeared in the witness' credibility. It concerned entries in the Hospital records relating to S.M. Wakefield when he was released from arrest in a state of mental distress. Whether the Court fully understood what all this was about is doubtful but the suggestion that Turnbull had cooked the books to injure Crawley was successfully planted in their minds.

Among the large number who came to testify to Lilley's exemplary character and sobriety was Colonel Shute.

As he was about to step down, Crawley asked permission to cross-examine him on a letter of censure on the state of the Regiment that had been sent by the Horse Guards on 30 June, 1860.*

Horsford submitted that the witness should not be allowed to answer any questions unless the letter was placed in his hands. But Shute said he was ready to give any replies which would make the letter unnecessary, adding that *he* was not on trial.

Crawley, whose obvious aim was to have the letter produced as documentary evidence of the state of the Regiment before he took over, and had no desire for an unscripted discussion with its former commander, declined the offer.

He was allowed to apply for it to the Horse Guards. And though his request was refused, he had notched up yet another point. *The Times* noted that although he regarded the letter as 'in the last degree essential to his defence', the Horse Guards had decided not to let him have it on the grounds that it only referred to 'personal bickerings of an old date'.

---

* The letter (which now seems to have disappeared) must have resulted from the stringent comments by Sir Hugh Rose after his Inspection the previous March. The reproof was probably mild enough. Sir Henry Somerset, the C.-in-C. had effectively removed the sting from Sir Hugh's remarks when he wrote in his covering letter that the Regiment 'is generally much to be commended and it is in excellent and serviceable order'.

25

The Drunkenness charge

Although the Court had decided that Lilley's character was immaterial to the Charges, it failed to check Crawley's persistent efforts to prove the dead man had been a drunkard. This he evidently thought necessary on two counts: to show Lilley had been unfit for his high rank and to counter the accusation that he (Crawley) had trumped up the charge in the first place to save his skin.

On the evidence he can hardly be said to have succeeded. Most of the witnesses from all ranks spoke of their respect for the dead man and said they knew nothing of his being addicted to drink. Two only testified that he had drunk to excess before his arrest. The most outspoken of these, Sergeant Mills, the 'Prosecution witness' who had unexpectedly revealed himself as a staunch Crawleyite to the discomfiture of Horsford, gives the impression of paying off old scores.

It is, however, a measure of Crawley's advocacy not to mention the Court's inconsistency, that Mills was later recalled to specify the occasions when he saw Lilley drunk. He said:

'In the first place it was in the year 1844 at Mr Daniel's the *Three Crowns* in Parliament Street, Nottingham. I saw him again drunk at Mr Wilkinson's the *Victoria* at Newcastle-upon-Tyne. I saw him again drunk at Balaclava; he was in the company of S.M. Moreton; he sent me down for three bottles of gin and gave me a sovereign to pay for them.*

* A rumour that Lilley was drunk at Balaclava and had missed the 'Charge' for this reason had been circulating in England. It was scotched by a letter to *The Times* in

179

'I saw him again drunk at Scutari. I pulled off his boots and covered him up in bed that night. I saw him drunk in Durham, when he was on the recruiting service, when the squadron lay there. I found him on his bed drunk.

'I also saw him in the female hospital which he occupied when we first marched into Mhow; he lived there. The day we marched into Mhow I was regimental orderly sergeant. I went up to my tea at 7 o'clock, and he ordered me away like a dog; he was then drunk, and I went away without anything. At Burhampore I saw him drunk in his tent on the line of march.'

Mills said that no one else to his knowledge saw the events he described. (Moreton, the only other soldier named, was, of course, safely away in India and could not be questioned.)

Another N.C.O. said that Lilley took 'arrack and coffee' regularly and had drunk 7 or 8 glasses of brandy and soda in one morning. But asked how he knew this, he said, only from Lilley's 'own confession'.

On the R.S.M.'s drinking while he was under arrest, one sentry described how on the day before his death Lilley pulled a bottle of brandy out of a long boot under his bed and offered him a drink. He refused, but later disposed of the bottle at Lilley's request.

Asked by Horsford if Lilley gave any reasons for drinking, the sentry said—'To the best of my recollection he said he was depressed in spirits.'

The sentry on duty when Lilley had his seizure told how he tried, at Mrs Lilley's request, to give him brandy with a teaspoon from a glass but was unable to get much down his throat because his teeth were clenched tight. There was continued questioning as to the amount actually swallowed, apparently to suggest that Lilley's craving for drink was insatiable even when he was dying.

Barnett was eventually recalled to give a more precise account of his conversation with Mrs Lilley on her husband's drinking habits.

He was first asked by the Court whether he thought four or

October from his brother who quoted Army Records to show that Lilley did not arrive in the Crimea until a year after the Battle took place.

five glasses of brandy a day was the 'usual medical comfort' for a woman in consumption during the hot season at Mhow. He replied that it was not an unusual quantity in view of Mrs Lilley's extreme exhaustion, as 'the great irritation from her stomach and bowels from which she constantly suffered did not admit her taking solid food or nutritious broths'.*

He said that when Crawley spoke to him of Lilley's having drunk to excess, he first consulted Dr Turnbull then went to Mrs Lilley and asked her whether it was true or not. She replied that her husband did take some brandy and soda or brandy and water every day, adding, 'but I do not think so much as to do him harm. Do you think it did, Sir?'

Barnett said he told her he hoped not. That was all. She did not mention any particular quantity. He wrote the Addendum because Crawley had stated so emphatically that Lilley had taken a large quantity of brandy, from which he inferred that the R.S.M. had drunk it.

The Court: 'Is it the practice in the 6th Dragoons to frame medical reports upon the statements of the Commanding Officer?' *Barnett*: 'No, it is not.'

Crawley then asked whether he meant to say that because he had been told the quantities of drink the deceased *had*, he took no steps to find out how much he had *drunk*? Barnett said he did not. It was rather a delicate subject on which to question a sick woman about her dead husband.

At the end of his interrogation, he asked for permission to add that he now recollected that at Mrs Lilley's request Mrs Cotton produced a bill of the Parsee's in which the brandy was charged and which he looked over at the time.

---

* This diagnosis does not seem to have been questioned by any medical authority at the time.

26

The Case for the Defence

So successfully had Crawley's supporters infiltrated the ranks of his opponents that there was a degree of truth as well as perhaps unintentional irony in the claim with which he opened his formal defence. The witnesses for the Prosecution, he said, had so effectively refuted the charges against him that there was no need to trouble the Court with further evidence. He would confine himself to inserting a few links and bringing forward some documents that were necessary for a proper understanding of the case.

He then produced the letter in which Sir Hugh Rose defended himself and his subordinates against the accusations made in the British Parliament and Press.* From it he read a passage where the General referred to *The History of the Inniskilling Dragoons* he was compiling. This document, he declared, was essential for his defence and he called on the Prosecution to produce it.

Once again he had caught his opponents unprepared. Horsford denied all knowledge of the letter and said it must have been intended for the private information of the Commander-in-Chief.

Crawley persisted, and the Court, obviously impressed by the illustrious name of the *History*'s author, decided to let him apply to the Horse Guards for it.

Meanwhile he asked permission to read out the whole of Sir Hugh's letter. Horsford again protested. He said that although he would allow any of the documents previously handed to the

* His main points are outlined in Chapter 17.

Prisoner to be read out subject to legal objection, the Court should see the originals first to decide whether they ought to be read out in public.

No, said Crawley. The proper course was for him to read them in open court, after which the Prosecution could raise any objections they pleased.

The President agreed and the letter was read.

Horsford again objected. Such a letter could not possibly be regarded as legal evidence for it merely contained opinions which depended entirely on hearsay. Sir Hugh Rose had not been present at any of the events he referred to, he was not speaking on oath and could not be cross-examined on anything he had written. However, now the letter had been read in open court, it was futile to treat it as inadmissible. The only course was to append it to the proceedings to be treated by either party according to 'its seeming worth'.

He then appealed again for the other documents to be read first in private to prevent what was bound to happen in the case of Sir Hugh's letter—their being published in the newspapers even before the Court had decided whether they were admissible as evidence.

Crawley said pointedly that he had no objection to a letter from Sir Hugh Rose being treated according to 'its seeming worth', but documents had been put in evidence against him (Sir William Mansfield's letter on the arrest of the Sergeant-Majors) without private investigation. Why was the Prosecution seeking to turn an open court of justice into a secret inquiry? He had been given a list of papers by Sir William Mansfield expressly to use at his trial and the Prosecution had allowed him to take copies. Under no circumstances would he consent to the *private investigation* they proposed.

Horsford insisted that in Courts of Equity, judges always scrutinized documents to decide if they were admissible before allowing them to be used in the proceedings on which judgements would be formed. The same applied to courts-martial.

The Court closed. It returned to say that the Prisoner could read out the documents.

The enemy in retreat, Crawley consolidated by asking to be allowed to read out *all* the documents together after which the

Prosecutor could object to any parts and state his grounds for doing so collectively.

This too was granted. All the documents giving the views of the Indian Authorities were read, making the high-ranking officers concerned yet more firmly established as sureties for the Prisoner's conduct.

Of the few witnesses called for the Defence, Captain Curtis, who had just joined the Regiment and was living in Crawley's house when the events at Mhow came to a climax, emerged as the most important.

He described dramatically what happened when Lilley made his complaint about the sentries at the Court-Martial. He was sitting next to Crawley and remembered vividly the look of astonishment on his face as he turned and exclaimed, 'Good God! can this be true?'

He also gave a graphic account of their departure from Mhow. About 9 o'clock on the morning they were to leave for England, a colonel of a native infantry regiment came to tell Crawley that the men of the Inniskillings had turned out on the road and were waiting to see him off. Then:

'As we came up towards the church we could see the men running from the barracks to get on to the road so as to be in time to meet the colonel. On our coming up they gave three or four hearty cheers and said, "God bless you, colonel. We hope you will return soon. Good luck to you." The colonel thanked them for their good wishes and drove on. About twenty yards further on the women of the regiment had turned out. The greater portion were in tears. They also expressed their good wishes to the colonel, and we drove away.'

Probably the most chancy move by the Defence was to call Ardaseer Framgee, the Parsee merchant who supplied the Lilleys with drink and other supplies. After the bills had been discussed, he put in a statement to the effect that Lilley used to call at his shop daily to drink four to six glasses of brandy for which he never paid. This he explained was because the R.S.M. ordered large quantities of everything for the officers' and sergeants' messes.

Cross-examined by Horsford, who at last seems to have

realized the extent to which he had been misled by the Court's assurance that Lilley's character had nothing to do with the Charges, the Merchant described how he was asked by Davies to go to Crawley's bungalow on 1 July to write out a copy of the bill.

In answer to further questions he said that, after his arrest, Lilley continued to obtain supplies from him by sending his *gorrawallah* (horse-keeper) to the shop. He personally gave this servant some of the brandy, but most of the goods were supplied by his assistant who was not in England.*

A pass-book was kept by Lilley to record the items purchased and Mrs Lilley continued to use it after her husband's death. He did not know what had become of the pass-book.

He bought some of the dead man's things from Mrs Lilley through S.M. Cotton and paid her what was left of the money after deducting the amount of the bill. But he could not remember if he paid the money into her hands or whether he receipted the bill.

Horsford, perhaps elated by his successful cross-examination proceeded to sum up;

The articles, he said, were supplied mainly by Ardaseer's clerk who had not been sent to England.

A pass-book was used when the goods were ordered. This had not been produced nor had any reason for not doing so been given by the Defence.

The articles were not supplied directly to Lilley or his wife but to a native horse-keeper, who had not been sent to England.

That is, there was no proof the articles were ever ordered by the Lilleys or that they ever reached them.

In reply to this impeccable reasoning Crawley raised the objection that the Prosecutor should have reserved his comments for his closing speech. He was supported by the President.

Cross-examined further the merchant admitted having conversations with Dr Barnett and S.M. Cotton over the bills. But when he was asked to relate what he said, Crawley again objected on the grounds that the conversations had taken place between the witness and a third party, *not in his presence*. They should not be admitted as evidence.

* He refused to embark. Page 147.

So began the last of the long legal skirmishes and the last of the somersaults performed by the Court over the relevance of Lilley's character to the charges. Horsford, now fully alive to the importance of this issue, said he wanted to recall S.M. and Mrs Cotton and Dr Barnett in order to go fully into the question of the bills. Crawley strongly objected and the argument dragged on. Then the President announced that the Court had decided the matter was not relevant but if the Prosecution wished to test the legality of his decision he could put it to the Judge Advocate-General.

Horsford said he would accept the court's ruling. Then, with the last kick of a drowning man, he said he would like to ask one question of Ardaseer Framgee.

Crawley demanded to know what the question was.

Horsford said he would put it in writing.

No, said Crawley. The prosecution and Defence were both closed. If that witness was recalled and questioned, he would have to re-open his Defence and all the witnesses would have to be re-examined.

Horsford said he would not press for the merchant's recall.

The Prisoner produced thirty witnesses to his character ranging from Generals to subalterns, and covering the period from Sandhurst to his taking command of the Inniskillings. Some had served with him for long periods.

Typical of the sentiments expressed was the concluding sentence of the testimonial from Sir George Russell Clerk, KCB:

'Such being the real character of Lieutenant-Colonel Crawley, he is, of course, incapable of an act of cruelty, malice, or any other meanness towards man or woman, in a regiment or out of it.'

27

The Closing Speech for the Defence

While the Defence's closing speech must have owed much to
the professional skill of Vernon Harcourt, the style and tone are
unmistakably those of Crawley himself.

It began with a rhetorical introduction in which he appealed
to the traditional attitudes and prejudices of his judges,
continuing the process of identifying them with him in his
predicament that had gone on throughout the trial.

It might be supposed, he said, that the day an officer who had
the honour to command a regiment second to none in Her
Majesty's Service was arraigned before such a tribunal would
be one of bitter humiliation. Yet for him, the prospect of this
trial had been his only comfort for many a weary month.

Unlike his unseen enemies, those anonymous writers who
were using him as an instrument to attack the whole British
Army, *he* had no distrust of the tribunal before which he now
stood. He was an old soldier and knew courts-martial well.
From them the guilty man had most to fear and the innocent
most to welcome. A soldier required no better than to be tried
by his peers, and he pleaded at last with inexpressible relief before a tribunal which knew by habit what were the duties of an
officer and by instinct what were the feelings of a gentleman.

At the end of his long preamble he said he would turn to the
charges. But first he hoped the Court would allow him to note
some of the offences he had been publicly accused of. It was said
he had used harsh language and had been overbearing towards
his officers; that he had perverted justice, intimidated witnesses

and suborned evidence at the Mhow Court-Martial. But as the charges made no mention of these, the military authorities must be satisfied they were all without foundation.

It would indeed be fatal if officers serving their Queen in distant lands had to look for fortunes or disgrace, not to their consciences and the orders of their superiors, but to the opinions of ill-informed and irresponsible writers. For then no man in authority would dare to do his duty.

Men like Sir Hugh Rose and Sir William Mansfield had seen one mutiny and knew how it began by condoning the insubordination of non-commissioned officers. They knew that early severity was the truest mercy. For though close arrest might be severe, 'it was better than blowing your soldiers from guns'.

Had he been allowed he would have proved by the letter from the Horse Guards to Colonel Shute and the *Historical Memoir* by Sir Hugh Rose, that at the time of the Court-Martial the 6th Dragoons was in a dangerous state of insubordination. But even without them he could satisfy the Court, 'who were all officers and gentlemen and were, or had been, commanding officers'. For they had seen how some of his officers conducted themselves in court and could well imagine what his position must have been on parade and in the mess-room.

Insubordinate officers naturally sought allies amongst the N.C.O.s. The Sergeant-Majors had been enlisted by the cabal to overthrow his authority . . . 'the crisis had arrived, the danger was imminent; one more step and the flame might have been kindled in the ranks and the famous Inniskilling Dragoons would have broken into open mutiny'.

But he knew his duty. He took the evidence and placed it before his superiors. From that moment he deferred to their judgement, which he unfeignedly knew to be better than his own, and acted in strict accordance with their directions, not only from a sense of duty but with a profound confidence in their superior wisdom.

The root question was one of mutiny or no mutiny. Had these men been tried, the Court would know that neither he *nor his superiors* had done anything not consistent with justice, discipline and the maintenance of the British Army—and the BRITISH EMPIRE.

He would now deal with the specific charges.

The legality of the arrest was conceded so he was justified in assuming the guilt of the Sergeant-Majors. Close arrest was necessarily severe. So what was the Prosecution putting forward? Was it that a man of plethoric habits, who happened to be married and whose wife was sick could not be subjected to the same military discipline as others? If so, the sooner such men left the Army the better.

It was inevitable that in the execution of the same sentence, military or civil, some men were bound to suffer more than others. This case had been treated as one of peculiar military barbarity; yet in what way was it different from a civil one?

Suppose that Lilley had not been a soldier guilty of a grave military offence,* but an unfortunate civilian unable to pay his debts; that the bed was sold under his wife who was then carried off to die in a work-house; and that he was confined in a debtor's gaol, where, under the combined effects of his sedentary habits and his predisposition to congestion, he fell sick and died. Would the judge who made the order and the sheriff who carried it out then be put on trial?

When the Sergeant-Majors were arrested, they were all placed under the same restraint. Yet the cases of Wakefield and Duval were not called in question. This was strange because Lilley had not been kept in solitary confinement like the others. Knowing the circumstances and not wishing to inflict upon the R.S.M. the additional punishment of separation from his wife, he had not sent her to the female hospital. He would leave the Court to explain the juridical logic of his being charged with excessive severity towards the one who had been treated with the greatest leniency.

It was a waste of time discussing the first quarters occupied by Lilley. And even Major Swindley admitted the second was as good as the one in which his sergeant-major was housed.† He was sure many officers present, like Sir Hugh Rose, would have been glad of such accommodation.

On the question of hardship and indignity suffered by Mrs Lilley, 'the evidence was all one way'. As proof of this he had

* He was not, of course.
† Swindley's sergeant-major was a single man.

only to draw the Court's attention to the men who had given evidence.

'. . . Sir, do you suppose these fine fellows do not think the suggestion of indignity to Mrs Lilley an imputation on their honour no less than mine? Is there any man here present who would not trust the charge of a sick and suffering woman to the manly good sense and high-bred feelings of the youngest private of the 6th Dragoons, rather than leave it to the tender mercies of the delicacy of Mr FitzSimon? . . . Where is the sentry who was within two feet of Mrs Lilley's bed? Why have the Prosecution not produced him? They could not find him because he never existed.'

The Prosecution had tried to retrieve the fortunes of a collapsing case by dragging an unfortunate defaulter named Gaffney from his prison. But he had only testified to his shame in the hope he might contribute to his Commanding Officer's disgrace.

Turning to FitzSimon's letter of complaint, Crawley said he had written his comments on it at the time and had wanted them forwarded to the C.-in-C. But General Farrell, out of consideration for Mr FitzSimon, had advised him to withdraw his letter and anyone might have expected that to be the end of the matter.

But what did Mr FitzSimon do?

'He formally withdraws his appeal to the Commander-in-Chief in India and so gets rid of my comments on it. The next we hear of the joint composition of the late Adjutant and the Quartermaster is that it has it has been forwarded by Mr Smales to the Horse Guards.

'Of course Mr FitzSimon has not the faintest idea how the document came into the hands of Mr Smales, or why Mr Smales sent it to the Horse Guards. These are exactly the sort of circumstances which are sure to have escaped the somewhat abnormal memory of Mr FitzSimon; and so by this very pretty device he contrives to blast my character without the disadvantage of having his statement confronted with my reply.

'If Mr FitzSimon's letter had been proceeded with, it would have been he, and not I, who would have been placed on trial. But by this ingenious manoeuvre he has contrived to reverse the situation and has managed to place me on trial and to shut my mouth whilst he gives evidence against me.

'I don't know, Sir, what the Court may think of such treatment as this of a commanding-officer by an officer under his command, but such is the origin of the second charge; for it is upon that letter so withdrawn before my face and so reproduced behind my back, that the second charge is founded.'

The charge itself was difficult to deal with because it was now proved to be based on a double untruth—that Mrs Lilley's privacy was intruded upon and that this was done by his orders. Mr FitzSimon had confessed that the execution of the order to post the sentries was in the hands of the Adjutant or the R.S.M. Yet neither he nor the Acting R.S.M. did see to the posting. When pressed, Mr FitzSimon said, 'I should say', that was Mr FitzSimon's favourite phrase, 'I should say it must be left at the discretion of the sentry'. Was any commanding-officer's reputation safe in the hands of such an Adjutant?

(While FitzSimon came in for the greatest flow of invective, other officers received their share in proportion to their importance as witnesses.)

'Major Swindley is a witness of another stamp, though not of a dissimilar animus. I don't know, Sir, what you may think of the major of the 6th Dragoons, who draws a nice distinction between his private and his public hostility to his colonel. Whether he has observed that fine discrimination on which I suppose may be considered a public occasion, it is for you to judge. He too, forgets for the moment the concerted oath, but he comes back after luncheon to amend his evidence, and to repair his oversight by giving out the 'damn' with an emphasis which will not easily be forgotten.

'Sergeant-Major Cotton's memory is less tenacious, probably because his spirit is less hostile . . . And then, Sir, there is Mr Wooden, the accomplished letter writer of excellent memory. He says he never recollects having had a bad memory. I fear this very assertion convicts him of the weakness it is intended to deny. He seems to have forgotten an occasion on which his memory appears to have been much less copious than at present. Look at his letter written on 3 October, 1862, and compare it with his evidence given on 23 November, 1863, and I think, Sir, you will be of opinion that the excellent memory of Mr Wooden consists principally in its progressive character, and that, like full-bodied wine, it vastly improves by keeping. It is fortunate for me that this trial has not been postponed for another year; for it is impossible to conjecture, now that Mr Wooden has got rid of his notes, what additional particulars he might not by that time have remembered.'

The fact was, the charge rested on the unsupported, or rather contradicted, evidence of Mr FitzSimon. Having heard and seen him the Court would ask themselves whether the reputation of any man living ought to suffer by the testimony of Mr FitzSimon.

And now he had disposed of the charge, what did they really think of the case. Did they not believe that his enemies had 'used the deathbed of Sergeant-Major Lilley to inflame popular prejudice, to prevent public justice, and make it the lever of a foul conspiracy to ruin the man they desired to destroy'?

If any hardship had been inflicted on the Sergeant-Major, would not the Regiment have been in arms against their Colonel? But their attitude towards him was shown clearly by Captain Curtis' account of the send-off he had been given when he left India under arrest.

'I will venture to say you have not forgotten the story of my departure from Mhow. What a spectacle. The murderer of Sergeant-Major Lilley is pursued by the cheers and blessings of the brothers-in-arms of the injured man. They bid him Godspeed in his hour of trial! they pray for the return of this harsh, wicked, unjust tyrannical Colonel. . . .'

But the malice of his enemies had supplied him with another testimony. He knew the wife of Sergeant-Major Lilley. A better woman never existed. If what he was accused of were true, would not her dying breath have testified against her husband's oppressor?

But what did Mrs Lilley say? The Court must have seen her last letter. He could not put it in evidence. But he thanked God his enemies had produced it for him.* The letter they had used to inflame public opinion against him was one which no fair, honest, right-judging man could read and not see in it the conclusive evidence of his guiltlessness of this awful crime . . . 'if this unhappy man died under my oppression, his more unhappy

* See page 152.

wife did not, could not, would not have written that letter.

'The law has always attributed a peculiar sanctity to evidence given in the article of death. My enemies tender you such evidence; it wants the sanctity of no oath, it is sealed by the awfulness of impending judgment. . . .'

He acknowledged the fairness of the Prosecution, which was only what he expected from the gentlemen who conducted it. He had fought a battle which was not his alone but that of everyone who might be similarly placed. He had fought with no friend but truth and no protector but the All-knowing and the All-mighty. His enemies had sought his shame and accomplished their own. He would not deny them the satisfaction of knowing they had partly succeeded, for the cost of the tremendous struggle would exhaust his slender fortune, the only provision for those dearer to him than himself. But though they had ruined his fortune, his reputation was beyond their reach.

He concluded: '. . . I know what your sentence must be because it will be in accordance with the evidence and the truth. I cannot hope that you will in every detail consider my conduct faultless. I cannot expect that in the long course of these trying events you may not discover in my behaviour, under unexampled difficulties, some errors of judgement and some defects of temper. I can only say, let those without such sins cast at me the first stone.

'But of these grave and serious charges I know you must absolve me, for I am innocent of them. You will give me back my character. You will give me back my sword—a sword which has been for thirty years—aye and which shall be again—at the service of a nation at whose hands, in the language of the charge, I have suffered "great and grievous hardships". You will redeem what remains to me of life from dishonour, and you will rescue my memory from disgrace. Sir, I await the sentence of the Court.'

28

The Prosecution's Reply

Throughout the trial the uncommitted public had veered in its opinion from day to day as the conflicting evidence unfolded. But following the closing speech on 17 December, many people made up their minds that except for the formalities, Crawley was home and dry. So strong indeed was the effect of his high pitched oratory that some even expected the Prosecution to waive its right to reply.

But in this at least they had miscalculated. When Horsford rose the following day it became clear that far from conceding defeat, he was about to make a strong if belated bid for victory by presenting the evidence against the Prisoner with a force, logic and clarity that contrasted with his earlier mishandling of the case.

The explanation for the change is obvious. Mr Denison, realizing that to leave the Closing Speech to Horsford would be to invite annihilation, had decided to give it the professional touch by writing it himself.

It was, of course, read by the Prosecutor, although there were rumours later that he refused to read certain parts. And at a guess this may have been true, for as a military man he may well have been inhibited from pitching it too strong against the Indian Command.

He began by regretting the mis-statements made in print regarding the events leading to the Trial, which had not only harmed the Prisoner but made the administration of justice more difficult.

He then outlined the essential facts about the arrest of the Sergeant-Majors, pointing out that there was a difference in the alleged grounds for holding Lilley and Duval from that for Wakefield who was said to have uttered threats of poisoning or otherwise doing away with anyone who gave information about the reading of Smales' Defence. For while there might be some pretext for keeping Wakefield in close arrest, there was none for Lilley or Duval.

When a commanding officer was entrusted with the arrest of an R.S.M. in the conditions that existed at Mhow, he might surely be expected to employ the least severe form of restraint consistent with the safe custody of his prisoner. The course of the arrest of R.S.M. Lilley showed how the Prisoner used the discretionary powers entrusted to him.

On the day of the arrest, 26 April, 1862, orders were given for the sentries to be placed outside the R.S.M.'s quarters.

On 28 April the sentries were placed inside the quarters.

On or before 4 May the sentries were posted inside the same room as the prisoner with orders not to lose sight of him day or night.

The proof that Colonel Crawley ordered the sentries to be placed inside the quarters was provided by his own letter of 4 June, addressed to the Deputy Adjutant-General, Mhow Division, in which he wrote that ' . . . it was deemed advisable to place the sentries in the rooms with the prisoners, not only to avoid the exposure of the sentries to the sun and hot winds, but to prevent the witnesses from being tampered with during the remainder of the trial.'

The Prisoner removed all ambiguity from the term 'tampered with' when he described it in the same letter as 'being communicated with in the interests of Mr Smales'.

He also wrote that on 25 May, immediately the necessity for close arrest ceased to exist, he had asked for it to be changed to open arrest. Yet, the fact was, Smales had already closed the examination of his witnesses on 10 May.

It was thus shown from the Prisoner's own written documents and the proceedings of the Mhow Court-Martial that for the 15 days from 10 to 25 May, a sentry was kept inside the Sergeant-Majors' quarters without any reason, not even the illegal one assigned at first. Far from being a merit, as the Prisoner tried to claim, his request to have the nature of the arrest changed was a condemnation of his conduct in that he ought to have made it 14 days before.

Horsford turned next to the severity of ordering the sentry not to lose sight of his prisoner day or night. The existence of the order was proved by its being passed from sentry to sentry, by Sgt Mills (called by the Prosecution, but a zealous witness for the Defence) who said it was part of the sentries orders, and by Lt Snell, whose evidence showed the order to be in force up to 7 May, when it was modified to the extent that the sentry was told not to go into Mrs Lilley's bedroom.

That the order was given by the Prisoner in the orderly room on or before 4 May was proved by the testimony of seven witnesses of various ranks and the fact that their evidence varied a little showed it had not been concocted. Nor had it been shaken by the Defence.

As to the annoyance inflicted on Mrs Lilley by the order, the Prisoner seemed to think that this could be measured by the opinions of the sentries and that to understand her feelings all that was necessary was to find out theirs. But how could the Court be expected to believe that a 'chick' would have prevented vexation to a dying woman—'a vexation that did not depend on the mere fact of a sentry seeing or hearing her but on the fixed ever-present knowledge in her mind that there was about her bed, about her path and spying out all her ways a constant, shifting, watching male stranger, who seemed to be by law, not only entitled always to be there, but whose duty it was to be so. Surely so flimsy an argument is as easily seen through as the chick itself.'

No need for the evidence of Pte Gaffney, whom the Prisoner did not attempt to cross-examine, and no witness had been called to contradict; who, though a Defence witness, was made the subject of a formal protest when called by the Prosecution. He was a prisoner for drunkenness. But there were worse crimes. His sobriety may have been defective, but he had too much delicacy to intrude upon a dying woman and too much manliness to flinch from telling the truth.

And what of Atkins, the old soldier of twenty years' standing, who had known Lilley since he joined. Atkins had said that Mrs Lilley did not mind him because he was a married man. The Court could judge how many sentries were married and so estimate how much freedom from annoyance had been secured by Mrs Lilley. For this conversation took place in the first bungalow which, compared with the second, was a sort of palace. They could think too of the freedom from annoyance when Mrs Gibson spoke to her in passing and Pte Little was confined for it.

Lilley made his complaint on 7 May. Yet except for the sentry being told not to go into Mrs Lilley's bedroom, the same close watch was still kept up. Why? And why was it continued after Smales closed his case on 10 May? The true explanation was that the Prisoner used 'his discriminatory powers to gratify his feelings of anger and dislike against the Sergeant Major'.

Mrs Lilley had gone out whenever she was able and since she was not searched, could have passed messages to her husband. The severe conditions of the arrest were therefore nugatory except for the annoyance to Lilley and his wife and for ensuring they were of 'a positively penal character'.

12 May, the date Lilley was moved to the second quarters should be carefully noted. It was two days after Smales had examined his last witness and about the time Sir William Mansfield's letter arrived to say there were no legal grounds for holding the Sergeant-Majors.

Yet the Prisoner, who had just expressed deep concern at the inconvenience to Lilley's sick wife, was shown by the evidence of Q.M. Wooden to have refused to allow him to be given either of the more suitable quarters available—one because it was too near the soldiers of the Regiment and the other because it would have placed him 'in the very jaws of Mr Smales'.

The Prisoner's solicitude for the comfort of the Lilleys was clearly outweighed by his desire to prevent all communication between him and the rest of the Regiment.

Horsford now spoke of the attempts by the Defence to prove that the second quarters were superior to those occupied by most married soldiers and particularly stressed the way in which Champion had tried to mislead the Court. But, he concluded, it was useless to labour the point, 'for it was proved over and over that Sergeant-Major Lilley and his wife occupied only the large room and that the inner verandah or passage called a room, was occupied by the sentry'.

He turned next to Sir Hugh Rose's letter of 28 July, and it was here that he came nearest to retaliating in kind to the shafts of irony hurled by Crawley at his witnesses.*

'Who is there,' he asked, 'on hearing Sir Hugh Rose's letter read out, did not sympathize truly with the writer; for he had much to complain of and much to endure; and he bore himself bravely under obloquy which, on this matter at least, seems undeserved. For if Sir Hugh Rose had been in possession of the facts that have been proved before the Court, justice might perhaps have been done in India as well as it will surely be done in England; but the worst evil that can befall a man befell Sir Hugh Rose. He could not get at the truth! But how was Sir Hugh Rose misled and by whom? . . .'

When Sir Hugh heard of Mr Fortescue's speech in the Commons he sent a telegram to Sir William Mansfield asking for full information on the quarters occupied by Lilley and his wife. He received a reply which stated that the R.S.M. had first been confined in a bungalow containing three rooms, and second at the end of the barracks in two rooms with verandahs round two sides.

From this Sir Hugh inferred that if Lilley and his wife occupied only one room they did so from choice and concluded that Mr Fortescue had based his serious disapproval, 'and very justly' on the assumption that they had only one room at their disposal.

Now, said Horsford, Sir Hugh was only reporting the evidence he had received from his correspondents and he was clearly misinformed on both bungalows. For it was proved beyond doubt that in the second bungalow Lilley had the use of only one room and that the sentry always occupied one of the verandahs. Therefore by his own statement Sir Hugh agreed with Mr Fortescue that for the Prisoner to confine Lilley and his wife in one room called for serious disapproval.

---

* Whether such a counter-attack on this particular target was well calculated to impress the Court may be doubted. It was one thing for Crawley to mock his subordinates, but quite another for Horsford to ridicule so famous a soldier as Sir Hugh Rose.

He would not waste the Court's time in considering how Sir Hugh came to be misled. But it would be some consolation for them to reflect that while their proceedings did not have the telegraphic swiftness which seemed to characterize Indian justice, at least by having models before them instead of tele-grams they would not fall into the double error of assigning fewer rooms to the first bungalow and more to the second. Of all the misapplications of science, the newest and worst seemed to be that of doing justice by telegrams.

Horsford next remarked on the unnecessary restrictions im-posed on Lilley during his arrest. Any good he might have derived from his daily exercise was neutralized by the degrading manner in which he had to take it. At first he used to jog round the race-course with his escort. But later, to prevent conversa-tion, the accompanying Sergeant was ordered to follow 15 or 20 paces in the rear. On 18 May, seven days before his death, he was allowed to sit on the verandah, but the concession was granted only grudgingly by the Prisoner.

The Defence had tried hard to brand Lilley as a drunkard, but, asked Horsford, how could it help their case to show 'that a singularly sober soldier was driven by misery to drink'?

Turning to the second Charge, of placing the blame for the annoyance to Mrs Lilley on to Lieutenant FitzSimon, he pointed out that Crawley had said that no one could be more shocked than he was when he heard Lilley say in Court that his wife had been 'incommoded or annoyed by the precaution taken for his safe custody'. Yet, at the time, instead of questioning the R.S.M. on the subject to see if the complaint was justified he set out to cross-examine him about the alleged crime he was supposed to have committed in order to discredit him as a witness. How could this course by the Prisoner be possibly reconciled with his statement that he was deeply shocked at hearing of how the sentry was placed.

The Prisoner had put forward the suggestion that when the order he gave resulted in annoyance to Mrs Lilley, it was FitzSimon's duty to alter it. But the members of the Court need hardly be told how absurd that was, or that it had nothing to do with the Adjutant's competency or incompetency. For an order so precise as that a sentry was not to lose sight of his prisoner by day or night left no room for a subordinate to consider what was best and fitting.

Yet the fact was that Lieutenant FitzSimon and S.M. Cotton *had* brought the mischief caused by the order to the Prisoner's attention in the presence of witnesses. His reaction had been to insist that it must be imperatively enforced or the Adjutant would be liable to all the consequences of disobedience. How strange then, with all these facts staring them in the face that the matter was even being considered. For if subordinates were allowed to qualify a precise order, the doctrine that obedience was the soldier's first duty would become null and void.

This was illustrated by the evidence of Cornet Snell, 'whose youth, manner and demeanour carried absolute sincerity'. When he made a change in the first order on the implication it was necessary, the Prisoner took no steps to ensure the amended order was a proper one. Instead, and on the very day he professed to be shocked at the effects of his first order, he cast all responsibility for changing it on a young officer who had never before performed the duties of Adjutant. As at all other times since he issued the first order on 28 April he refused to give his subordinates an order in writing and so left all the details of the arrests open to question.

Having dealt with the Charges, Horsford commented on the tactics employed by the Defence.

Whenever evidence prejudicial to their case was produced they immediately asked it to be considered as perjury. Because the Prosecution had not produced the sentry who stood in the manner complained of by Lilley, the Prisoner contended that his complaint was false and that he perjured himself in making it.

But how could this be reconciled with the Prisoner's own conduct when he took no steps at the time to find out whether the complaint was true or false and up to yesterday had allowed Lieutenant FitzSimon to remain under the imputation of inhumanity for an act which he now said was never committed. Was it not ridiculous to expect the Prosecution to find in November, 1863, what the Prisoner made no attempt to discover at the time when he professed to be so shocked in May, 1862.

The Prisoner said the second Charge 'rested on the un-

corroborated, nay contradictory testimony' of FitzSimon. But
that was not so. It was proved by the evidence of seven other
witnesses as well as S.M. Lilley.

The Prisoner attributed the mass of supposed perjury to the
influence of Mr Smales. But to support this allegation he had
produced not a scrap of evidence.

He had denied having any discretionary powers in carrying
out the orders for close-arrest, but the Court would now understand how he had abused them. For although he received Sir
William Mansfield's letter on 12 May telling him the Sergeant-
Majors could not be brought to trial for conspiracy, he treated
them as convicted criminals and actually asked the Court to
believe he was justified in doing so.

'Does not this,' Horsford concluded, 'furnish a clue to the
undue, unnecessary and utterly illegal severity of the Prisoner
towards these men?'

He would not weary the Court with comments on the
Prisoner's address, for a conflict of words was idle. He had taken
them through the evidence by a gate not known to the public and
by a way avoided by the Prisoner. It was left to them to decide
which way led to truth and justice.

29

Verdict and post mortem

Contrasting forecasts of the verdict appeared in the Press. On 23 December, *The Times* predicted the Court would announce its finding sooner than expected and that though the exact terms were a secret, it was pretty well understood that a favourable view had been taken of Colonel Crawley's conduct, making ample allowance for the circumstances in which he had been placed.

On the 24th, however, the London Correspondent of the *Grantham Journal* reported that in spite of the Defence's fervent appeal and before the Prosecutor had made his lucid and masterly reply, 'any amount of money could have been wagered at the professional and other clubs at the odds of 20 to 1 against the acquittal of the prisoner'.

Ironically, on the day this appeared in Lincolnshire, the London dailies carried the news all had been waiting for. Crawley had not only been given a full and honourable discharge, but for the second time had routed his enemies.

This was apparent from the comments appended to the verdict. 'The Court cannot close its proceedings without remarking on the animus displayed by Major Swindley when giving his evidence and on the evasive, hesitating and unsatisfactory character of the evidence of Surgeon Turnbull and Lieutenant and Adjutant FitzSimon and on the manner in which their evidence was given.'

The finding had to be approved by the C.-in-C., but as that was to take almost three weeks, there was ample time for comment by the Press. Most was approving or non-committal. But one aspect calls for special notice—the extraordinary conduct of the paper that had been so widely acclaimed for its humanitarian zeal in leading the outcry over the treatment of Lilley when the story burst upon the nation six months before.

Following the publication of J.O.'s letters and its highly emotional leader after the Parliamentary Debate, *The Times* had prudently retired into a studied impartiality and a policy of 'wait and see'. It passed little comment though numbers of letters representing all sides as well as extracts from other journals dealing with the questions raised appeared from time to time.

Then, on 24 December, with the verdict known, it came out with a leader, which, set beside that of 8 June must rank as one of the most remarkable and unashamed reversals of opinion in British journalism. A few extracts will make the point.

The trial, it said, proved that Lilley 'was actually living in a comfortable house, on a larger scale than suffices for many clergymen and half-pay officers in this country. . . .

'We would not utter a harsh or unfeeling word against Mrs Lilley, for whose character Colonel Crawley himself expresses a high respect; but truth is truth; and we cannot help saying that a lot of false delicacy has been wasted, in our opinion, on this part of the subject. Privacy is a relative term; a sergeant's wife is not accustomed to the same sort of privacy as a fine lady, and to say that Mrs Lilley's modesty was wantonly insulted because a sentry might have seen her lying in bed if he chose to look through the lining of a chick, or curtain, appears to us absurd. . . .

'We consider it clearly established that Colonel Crawley had no intention to persecute Lilley, but that no persecution was practised is the sufficient answer to the charge.'

The retraction of everything it had said six months before could perhaps be excused. But the leader's concluding remarks went much further:

'In the meantime it is for the authors of the cruel slanders which have cost the country this vast expense and their victim such unmerited obloquy to stand on their own defence; and for those who believed them to make the only reparation in their power by expressing without reservation, their sense of the injustice which has been done.'

J.O., upon whom this onslaught fell most heavily, refused to believe the case he had built with such impregnable logic had been defeated or that justice had been done. He accepted the challenge and wrote from Paris on 30 December stating with his usual clarity and style his reasons for disagreeing with the verdict. His letter was published, but not until a fortnight later. The next day another, signed 'Verax', supporting *The Times'* own views appeared, and to this J.O. was allowed no reply.

An exchange of letters between him and *The Times'* proprietor followed. After repeated refusals to have them published, J.O. had them printed privately. But his long, happy and fruitful relationship with the paper was over. For him the break was painful and lost him several old and valued friends.

The Times reported his death four years later with brutal brevity, avoiding all expressions of regret or eulogy.

Discussing the affair in his introduction to a selection of J.O.'s writings, Sir William Stirling Maxwell wrote of *The Times* having reversed its previous opinions, 'not only with a flourish of trumpets but also with a volley of musketry fired in the faces of those who had hitherto followed its lead'. He thought, however, that J.O. should have realized his position was hopeless once the tide of influential opinion began to turn in Crawley's favour. 'Observation should have taught him that timely retreat from any unpopular opinion was the habit and policy of *The Times.*'

By contrast, *The Guardian*, which had lauded *The Times* for its initial stand, stuck to its guns and remained unconvinced by Crawley's 'full and honourable acquittal'. It pointed out that in a country where English law was said to exist, Lilley had been confined by way of 'merited punishment, without the most distant imitation of a trial'.

Crawley had spent much time denouncing the anonymous writers who had maligned him. But when he came to considering the real points at issue his efforts were absurdly ineffectual. He asked the Court to rest its decision on the testimony of a witness whose competence could not be disputed, then proceeded to introduce, with abundant pathos, the deceased Mrs Lilley who had written a letter to her relations announcing her husband's death without attributing it to the ill-treatment he had undergone.

This letter did not, as Crawley suggested, have a peculiar sanctity in law, for it had been written with none of the formalities required for such a document. He should have been stopped by the Court as soon as he began to refer to it.

The *Spectator* had from the start been reasonably detached in its views on the affair. It now indulged in some lengthy philosophizing and decided that though 'Crawley's choleric remarks on minor infringements were not likely to make his influence extensive . . . the fact that his failure to discharge his duties caused the death of a good soldier was his [sic] misfortune rather than his fault.'

It went on to warn its readers not to allow their sympathies for Crawley to go too far, as this might provoke a counter-swing in favour of the Lilley family again. That would never do. For if it got abroad that an N.C.O. could suffer injustice causing his death without an inquiry, distrust would spread among the classes to whom the nation looked for its best soldiers.

It concluded with the noble sentiment that every shilling spent on the trial was well spent if it established the conviction in the heart of the nation that 'while public indignation may be appeased without a victim, no rank will be permitted to screen injustice'.

The *Lincolnshire Chronicle* also appealed for common sense by deploring the desire of some sections of the public to have someone punished for Lilley's death. Such people, it said, were unable to accept that he died by the Visitation of God. Because he was a man of good character they thought he should be immortal. But brandy and the Indian climate did not agree with the best constitutions, nor were they lenient to the best of characters. Close arrest it pointed out somewhat incongruously, did not kill Canning! Elgin! Wilson! or Dalhousie!

This journal also contradicted a rumour that the Earl of Cardigan was promoting a subscription in the Clubs to pay the heavy expenses incurred by Crawley in his defence.*

The *Illustrated London News* drew attention to Crawley's

---

* A fund was in fact set up. Headed by Sir George Brown, C.-in-C. Ireland, officers from Generals down paid in their hundred, fifty or twenty pounds. Probably some civilians too. One wrote to *The Times* to say he was ready with his guinea. At the end of the first day £500 had been subscribed.

comments on the Press. It admitted that he had suffered at the hands of the Press but that the Press could no more avoid inflicting suffering ' . . . once granted the Crawley Case was a fit matter for print—than Colonel Crawley, according to his own version could help inflicting suffering on Sergeant-Major Lilley —once granting the Sergeant-Major a fair subject for arrest.'

Confirmation of the Court's verdict came from the Horse Guards on 14 January. The Duke of Cambridge announced that Crawley would resume command of the 6th Dragoons with the least possible delay, and added the hope (no doubt fervently), that he could prove by his tact and judgement that he appreciated the importance of his position as Commanding Officer and that the painful experience of the past would not be lost on him.

As to the officers mentioned for their 'animus' both Major Swindley and Surgeon Turnbull had been similarly reflected upon by the Court at Mhow so that he could not continue to treat them with leniency. Swindley, whose feelings towards his C.O. struck at the root of all discipline would be removed from the Regiment. Turnbull's conduct over the Hospital Records would be further investigated, but whatever the result, he could no longer remain in the Inniskillings. Lieutenant FitzSimon had shown himself utterly unfit for the post of Adjutant and would also be removed.

The Duke deplored the 'tone and temper' of some of the officers who had given evidence, which revealed the amount of ill-feeling that actually existed and of which he had been un-aware when he issued his Memorandum on the Mhow Court-Martial.

He ended by saying that while he still believed Sergeant-Major Lilley's sobriety, up to his arrest, was supported by the evidence before the Court, he now understood how the statements made to Sir Hugh Rose, borne out by the Medical Officers, led that distinguished officer to pass the remarks he did after the Mhow Court-Martial.

This brief account of the post-mortem on the Crawley Trial, would not be complete without a reference to the debate that took place in the Commons on 15 March, 1864, the ostensible reason for which was a move by Dudley Fortescue to have certain papers connected with it laid before the House.*

Fortescue said he did not wish to call in question the verdict of the Court-Martial. There was a conflict of evidence and in accordance with well recognized principles of law, the accused had received the benefit of the most favourable construction. He had studied the proceedings probably more closely than any other member would have considered it worth his while but he was not called on to tell the House what opinion he had formed.

It would take years to restore confidence in military justice— if it ever existed. No acquittal or financial triumph on the part of one of the chief actors could do away with the recent public scandal. He only hoped it would rouse in the military authorities a firmer determination to do right and justice at whatever cost to individuals or apparent sacrifice of authority. But above all he hoped it would bring a review of courts-martial and 'some sweeping and radical change in those cumbrous and antiquated forms of procedure with their fatal temptations to perjury. . . .'

H. R. Grenfell, seconding, believed that if an earthquake were to swallow up the colonels, majors and captains tomorrow, the C.-in-C. would have no difficulty in replacing them. But if the same earthquake swallowed up the N.C.O.s he would have very great difficulty finding replacements for *them*.

He also drew attention to the treatment given to some officers who gave evidence at the Trial and hoped that the Government would make it clear that truth was what was required from witnesses 'not the favour of sergeant, colonel or even C.-in-C.'.

These views were opposed with rather more vehemence and less skill by the gallant and honourable members who attacked Fortescue for raising the matter in the first place and the Government for giving way to public opinion by having Crawley court-martialled.

But Lord Hartington walked the diplomatic tightrope with consumate skill and Headlam (the J.A.G.), gave a clear explanation of the steps he had taken to ensure that the trial was a fair

---

* The atmosphere was less emotionally explosive because of the absence of William Coningham with his 'impassioned utterances'. He had retired during the long recess after falling out with a body of his supporters on the subject of Temperance of which he was a strong advocate. He never returned.

one, claiming that, within the existing regulations he had done the best possible. He went on to outline the procedures in criminal trials in civil courts and suggested that, as far as possible, they should be the model for any reform in courts-martial.

By the end of the debate (80 columns of Hansard), with honour satisfied on all sides, it was resolved 'that the production of any further papers relating to the court-martial on Colonel Crawley is inexpedient'.

30

Old soldiers fade away

In any account of the subsequent careers of the participants in the Crawley Affair it is impossible to ignore at least three truisms—that there is a lack of vindictiveness in the British character enabling it to dispense justice without paying much attention to logic; that fate has its irony; and that old soldiers tend to fade away. Having mentioned this, the facts may be allowed to speak for themselves.

In the first flush of victory, Crawley seems to have toyed with the idea of bringing a suit for libel against J.O., but on advice thought better of it and left for India to disappear from the public gaze for the next three years.

The Inniskillings sailed for home in December, 1866, and it will surprise no one acquainted with their Commander's attention to minutae that horses, barracks and indeed everything handed over were in such tip-top condition as to evoke commendation and thanks from the C.O. of the in-coming Regiment. But strive as he may the perfectionist is forever prey to capricious fate and human frailty.

EXTRAORDINARY FREAK OF A COLONEL IN YORK

Under this intriguing title, the *Yorkshire Advertiser* gave a mock-heroic account of how the peaceful citizens of York were roused in the early hours of the morning 'by the strains of martial music and the clang and tramping as of troopers on the march'. Alarmed, they opened their windows to see the gallant Inniskillings marching past.

'A thousand surmises were indulged in, and those who entertained them caught the infection from the troopers' sulky faces, and finding none to relieve their doubts, retired once more to rest.

'On marched the troopers, led by their "very gallant" Colonel along the route . . . to the Hull road, along which they were marched some half dozen miles, when the order was given to wheel round and return homeward and to their quarters. These were reached at 5.30. The expedition was ended; but it was evident from the gloomy faces of the troopers, not at all improved by the long icicles that hung from their hirshute appendages that it had been of a most unsatisfactory character.

'Not to weary our readers further we will tell them all we know about this extraordinary march out. About the very small hours of the morning the Colonel was conducted to his barracks, and on entering therein, he found a "loose" horse on the square. This certainly was very wrong. The Colonel instantly ordered the trumpets to sound "Boots and Saddles".

'The regiment, gallant fellows as they are and ready for any emergency, promptly answered the call, and were no sooner assembled than they were marched out as described. It would thus appear that the whole regiment was punished for the carelessness of probably one man. The affair can hardly be hushed up by the War Office.'

In fact the news of this exploit appears to have caused little stir outside Yorkshire and Crawley soldiered on to December, 1868, when he retired on half-pay.* He was made a major-general in the usual way and lived in retirement for 12 years.

His death at the age of 61 at 9 York Terrace, Regent's Park, was probably quite sudden, for his will preceded it by only a few days. He left a personal estate of 'under £7,000' to be administered by his friend Frederick Waller, Q.C. (who had been his junior counsel at his Court-Martial).

Most of his effects and £250 cash went to his wife, who is not mentioned by name. His elder brother, Henry (who lived on into his 80s to survive him by more than twenty years), was left 'my watch and appendages, books, papers and manuscripts'. The rest was to be converted into a trust fund to pay his wife an

---

* The command of the Inniskillings went to Thesiger, whose quarrelling had set in motion the Crawley Affair.

annuity of £200 per annum, with anything left over to go to his nephew, Captain Thomas Crawley (the only member of the family still in the Service).

One has the impression he died lonely—the absence of children and the distant reference to his wife. There was a clause in his will leaving £50 to his coachman, 'if still in my service at my death'. Perhaps at the end he sensed the hollowness of his victories in the two courts-martial and that many people saw him as a charlatan, sheltered by an establishment protecting its own out of self-interest.

On the credit side, it can hardly be denied that he had fought with great resourcefulness and neglected none of his talents.

It would have taken many months for the news of his death to reach the individual whose 'upas-like shadow' had blighted his happy home almost twenty years back.

Although Smales had played no official part in the Aldershot Court-Martial, he had ardently hoped for a guilty verdict which would have done much to restore his own reputation. He must not only have been disappointed at the result but mortified at finding himself being portrayed as a sinister puppet-master manipulating the Prosecution witnesses to become a vital part in Defence strategy.

But while the Trial was in progress, his fortunes had taken a slight turn for the better. Perhaps from a faint desire to right a wrong but more from an overwhelming desire to rid itself (like Crawley at Mhow) of this persistent firebrand, the Military Hierarchy had been quietly debating the possibility of placing him on half-pay.

In this there was a problem, for Smales had sold his commission in 1848 and had not returned to the Regular Army until ten years later. Up to the date of his being cashiered, he could count only 3 years 10 months towards the five years needed for even the lowest pension. Had he continued to serve, however, the qualifying period would have been completed by 24 September, 1863, though there was no regulation that allowed the time since he was cashiered to be counted in spite of his having been pardoned.

In October the War Office submitted his case to the Law Officers of the Crown with a covering letter that hinted they would be happy if the decision went in Smales' favour. The hint was taken:

'We think it would be expedient and proper that her Majesty should be advised to place Mr Smales on half-pay at the rate of 6s per diem, being the rate to which he would be entitled if he had continued to serve as Paymaster down to the present time, and had now been incapacitated from further active service.

'The circumstances are very exceptional and such as to warrant, in our opinion, such an exercise of her Majesty's power. And we see no other way, under these circumstances, the fair requirements of justice can be reconciled with the interests of the public service.'

With the sanction of the Treasury, the War Office wrote to Smales to tell him he would be granted half-pay on condition he settled his accounts, and to this he agreed. A rumour that he had become insolvent in order to avoid paying the debts alleged to be outstanding in Regimental accounts under his paymastership was later quoted by Sir Hugh Rose in correspondence with the Horse Guards, but there seems to be no evidence of this and certainly his half-pay was never affected.

In 1866, however, at the age of 55, he decided that the climate in Britain was altogether too inhospitable, packed his bags and embarked with his family on the six months' voyage to a fresh start in Australia. It comes as no surprise that, indestructible as ever, he survived the rigours of pioneering for twenty-two years.

In his 78th year, his wife having died, he married again. A few days later, he too died—of heart failure.

By the time it was publicly announced in January 1864 that FitzSimon, Turnbull and Swindley, the three officers censured by the court, were to be removed from the Inniskillings, arrangements had already been made to place them on half-pay. Nor does the punishment seem to have seriously affected their subsequent careers.

Perhaps in the belief that FitzSimon had already suffered enough from his humiliations at the hands of Crawley and the censures heaped upon him, the Military Authorities pursued him no further and may even have attempted to make some amends. After four months he was allowed to transfer to the 62nd Foot

(1st Wiltshires). He also was given £490, the difference in the values of cavalry and infantry lieutenancies, to which, as he had purchased no part of his commission, he appears not to have been entitled. Five years later he was made musketry instructor, so presumably his eyesight was still excellent. He retired in 1878 on a major's pension and survived to the turn of the century.

Almost immediately after the Court-Martial's verdict, an investigation into Turnbull's Hospital Records took place. Predictably he was exonerated of any malpractice and appointed as Staff-surgeon. Barnett too was transferred, no doubt willingly, to the staff and did not serve again under Crawley.

Swindley remained 'unemployed' for two years and nine months, when he was appointed to the Cavalry Depot at Canterbury. Three years later he transferred to the 15th Hussars under Colonel FitzWygram and in 1874 was given the command of this Regiment in which Crawley had served so long. The end of his regulation period of office came in July, 1879, when he was taking part in the Afghan War. He was now 48, an age when the effects of his long service which included campaigns in Africa, the Crimea and India, might have been expected to have taken their toll. In fact he had completed little more than half his allotted span.

A few dates mark the fading process. In 1884 he was made an honorary major-general. In June, 1907, aged 76, he was nominated C.B. An entry in *The Times* of 21 March, 1913, announced his 82nd birthday. He lived on through the First World War and when death came in March, 1919, he was ten days short of 88.

At Mhow he had told Crawley he only attended Church when on duty. His last attendance was at Esher Parish Church, sixty years later.

What happened to Captain Weir is something of a mystery. He retired almost immediately after the Aldershot Court-Martial but the exact date is missing from the usual records. He does not appear in the Lists of Retired officers or in the Deaths.

Lieutenant Davies transferred to the infantry in 1870 and retired 10 years later with the rank of major. Soon after he was made lieutenant-colonel and was still alive in 1901.

Renshaw, whose youthful indiscretions had brought the conflict between Crawley and his officers to a climax, also retired as a major after having transferred to the 16th Lancers. His name was still on the Retired List when he was 77.

Of the two V.C.s, Malone continued as Riding Master until his death in South Africa with the Inniskillings in 1883. He was 50. Coincidentally this was also the age at which Q.M. Wooden met his death. He stayed a year with the Regiment, transferred to the 5th Lancers and then to the infantry. His end was tragic. He shot himself in barracks at Dover in April, 1876, 'while under temporary insanity'. As he was R.C., the phrase was probably not the conventional euphemism.

Colonels Shute and FitzWygram continued in public life long after the Crawley Affair was over, the solid respectability of their careers contrasting with the ludicrous and tragic events that developed out of petty squabbles during their stay in India.

When Shute left the Army in 1872 he put up for Parliament as a Conservative and was elected for Brighton (Coningham's old constituency). He held the seat for eight years. The robust constitution that carried him throughout the Crimean War without a day's absence from duty continued to suport him into a ripe old age. He died in 1904, aged 88.

FitzWygram became recognized as an authority on horses. He was active in forming the Veterinary School at Aldershot and in 1875 was elected president of the Royal College of Veterinary Surgeons. From 1885 to 1900, he sat in Parliament where he constantly sought to improve conditions of service in the Army. Claiming that he left Eton 'without the slightest knowledge of any subject which had been of the smallest use to him in after life', he urged that officers should be given a modern, practical education in place of the old academic one. He too died in 1904, a few months after his old C.O., aged 81.

Sir Hugh Rose continued as C.-in-C. India up to 1865. Eight months after Crawley's acquittal he sent the Horse Guards two long letters in which he continued to defend the rightness of all his decisions and also protested at the unjustness of the comments made on him personally by the Prosecutor at Aldershot. He complained particularly about remarks such as 'doing justice by telegrams', which, he said, contravened the accepted rule that anything approaching sarcasm should be carefully avoided in official pronouncements. He conveniently overlooked the numerous sarcastic comments by Crawley in his speeches at Mhow and Aldershot.

He denied, however, any desire to reopen the case, but asked that his letters should be placed with the other papers for posterity to pass judgement. The Horse Guards sent the unfortunate Colonel Horsford, who had read out the strictures on Sir Hugh's conduct, a copy of the letter with strict instructions not to answer it in any way.

In 1866 Sir Hugh was raised to the peerage as Lord Strathnairn, and eleven years later was made a field-marshal. In his declining years he spent much time examining religious questions of the day and denouncing atheism. He died at the age of 84.

The part played by Sir William Mansfield when, as Lord Sandhurst, he spoke for the abolition of the Purchase System in the Lords, compared with his conduct over the Mhow Court-Martial, has provided an interesting example of how much easier it is to be statesmanlike in a theoretical situation than it is to take administrative decisions in an emergency. Soon after he succeeded Sir Hugh Rose as C.-in-C., India, he demonstrated how little he had learned from the mistakes and misfortunes of the man he had supported, by becoming involved in a dispute with a member of his own personal staff and having him court-martialled for peculation and falsifying accounts.

Then, while the court was sitting, he made matters worse by appointing one of its members, an infantry officer, as Commandant of the Cavalry Corps. The scandal was discussed throughout the Army with regimental officers and men alike openly expressing the hope that the C.-in-C. would be defeated. Their wish was granted, but only partially. Though the officer was acquitted he was discharged the Service on disciplinary grounds.

Compared with most of the officers under consideration, Lord Sandhurst died young. He was 57.

Lieutenant Bennitt's evidence at the Mhow Court-Martial, one of the least savoury episodes of that outrageous Trial, naturally had no adverse effect on his career. By 1883 he had risen to second in command of the Inniskillings and two years later was given the command of the 5th Lancers. He was with that Regiment at Mhow in 1889 when a particularly virulent outbreak of enteric fever carried off fifty of its N.C.O.s and men.

He retired later that year to Slough where he became a J.P. One feels a natural curiosity as to what his attitude was to those appearing before him suspected of perjury.

He died on 21 March, 1931, aged 91 years and 4 months, probably the last living link with the Crawley Affair.

Appendix

Page 10 1. *The Purchase System.* Until 1871 officers bought and sold
their commissions. At regulation prices an infantry colonel
would have paid £4,500 for his rise through ensign, lieutenant, captain and major; a cavalry colonel—£6,175. A
'black market' existed that raised these prices to £7,000
and £14,000 respectively. Sir Henry Havelock wrote of his
'mortification' at being 'over-purchased' three times. Some
senior officers refused to retire until they received a 'purse'
from those who would benefit. Poor officers had to wait for
a brother officer to die for a 'free' step up or else volunteer
for an unhealthy foreign posting. Normally a 'free' promotion
could not be sold later.

Page 15 2. *George William Fredrick Charles, 2nd Duke of Cambridge.*
Only son of George III's seventh son, Adolphus Fredrick.
First cousin of Queen Victoria. Entered Army at 17. Com-
manded Division of Guards and Highlanders in the Crimea.
In action at Alma and Inkerman. Courage never in doubt
but was too impressed with horrors of battlefield for active
command. Wrote in diary after Alma, 'When all was over
I could not help crying like a child.' Invalided home.

Commander-in-Chief the Army in 1856. After creation of
Secretary of State for War lost administrative powers but
still represented the Crown and had some independence in
discipline, in appointments and promotion. After 1862 had
general control of troops in India. Relations with War
Ministers up to Cardwell (1868) were cordial but from
then on was gradually stripped of power.

Watchword—'Discipline, esprit de corps and the Regimental System.' Tended to regard Staff College Graduates

217

as careerists anxious to avoid regimental duties. Diary, kept from age of 14 expresses frank acknowledgement of his own shortcomings. Married an actress, Miss Fairbrother. Children named FitzGeorge.

Page 17   3. *Sir Hugh Henry Rose* (*1801–85*). Born and educated in Berlin. Entered the Army 1820–1830—Equerry to 1st Duke of Cambridge. Served in Ireland, Syria, the Crimea and India. Exceptionally brave and was said literally to know no fear. During the Battle of Inkerman rode through 'withering fire' reconnoitring Russian positions—the enemy were so struck with his courage that they ordered cease-fire. In Malta (1836), during an outbreak of cholera worked day and night assisting the doctor of his regiment to tend the sick.

Outstanding leader in the Indian Mutiny. In 1860 became C.-in-C. the Bombay Army and soon after C.-in-C. India. Carried out the amalgamation of the Queen's and East India Company Forces. Introduced regimental workshops and soldiers' gardens in cantonments. Issued an order that he would promote only on merit—all applications for appointments to be made through commanding officers, who had to report fully on the antecedents and qualifications of applicants. Was very severe on neglect of duty. Dismissed two brigadier-generals for not visiting hospitals during a cholera epidemic.

In spite of his undisputed qualities of courage and leadership, a rather different picture of the man emerges from his part in the Crawley Affair.

Page 18   4. *Joseph Malone V.C.* Born Eccles, Manchester, 1833. Joined 13th Light Dragoons 1851. Promoted from private to lance-sergeant. Appointed Riding Master in the Inniskillings when they left England for India in 1858. Married Captain Weir's daughter at Kirkee in 1860.

Wooden and Malone were each awarded the V.C. for similar acts of bravery—going to the assistance of a wounded officer after the Charge of the Light Brigade.

Charles Wooden V.C. Born London, 1827, of German parentage. Joined 17th Lancers in 1845. Appears to have retained his foreign accent—nicknamed 'T'ish me, the Devil' from a remark he made to a sentry who failed to recognize him. Had a ginger beard. Sergeant in the Crimea. R.S.M. in 1860 and transferred to the Inniskillings as Quartermaster the same year.

Page 26   5. The divorce, one of the first under the Matrimonial

Causes Act of 1857, was, in the words of Lord Campbell who heard it, characterized by 'circumstances of the most unusual profligacy'.

Mrs Renshaw was married to her first husband, John Joseph Tourle, a solicitor, in 1850, when he was 35 and she had just come of age, having been under the guardianship of a Mr Sterry, brother-in-law of Mr Tourle. Her inheritance of £2,500 was settled to her personal use together with two insurances of £700 and £300 on the life of her husband.

The marriage seems to have been happy enough for three or four years until they moved to Abbey Wood near Erith where they were introduced to 'all the families of position' round about, including a merchant named Renshaw who lived about half a mile from Ashmount-cottage, their new home.

In 1856, the Renshaw's eldest son, Richard William, who was at Cambridge, began to correspond with Mrs Tourle and became a regular visitor at the house when her husband was away. 'Improper familiarities' were observed by the servants and in January, 1857, Mr Tourle's suspicions were aroused by his wife's being away for some time in Brighton.

A reconciliation followed and the pair lived together until the following September when Mr Tourle wrote to his wife from Scotland where he was staying, telling her he had proof of her adultery with Mr Renshaw and had finally decided they must separate. At the same time he told her he would increase her annual income to £140.

Mrs Tourle replied assuring him of her 'unalterable affection'. But this time he remained adamant, and for a while she appeared to be quite satisfied with the financial arrangements made for her. Then in February, she decided to bring a suit for restoration of conjugal rights; at which the sorely tried patience of Mr Tourle became exhausted.

He had his wife watched, and, acting on information, went one morning with a friend named Morse to the London Hotel in Albemarle Street where they surprised Mrs Tourle with young Renshaw at breakfast.

'Some strong observations' were made by Mr Tourle after which he and Morse went into the communicating bedroom and found that the bed was not made. They also saw a box belonging to Mrs Tourle which contained the draft of a letter to her legal adviser on the subject of the proposed suit against her husband. This so infuriated Mr Tourle that he called Renshaw 'an infernal black-guard', and his friend prudently removed from his hand a stick he was carrying.

A good deal of evidence was produced to support the husband's petition.

John Sterry, his nephew, told how, after the move to Ashmount-cottage, Mrs Tourle 'was rather snappish with her husband, and on one occasion he went suddenly into the drawing-room and saw Mr Renshaw with his arm round Mrs Tourle's waist'. In cross-examination, however, he admitted that his mother (Mr Tourle's sister) had always disapproved of the marriage.

William Price, a lad in the Tourle's service, gave evidence of Renshaw's visits to Mrs Tourle when her husband was away. He also said he had taken notes to Renshaw from his mistress and brought answers back.

Alfred Peacock, a policeman, stated that late one night in January, 1857, he became suspicious of a man hanging round Ashmount-cottage and was about to arrest him when he produced a card saying he was Mr Renshaw. He then accompanied him back to his father's house for identification. On another occasion he saw Mrs Tourle with Renshaw in a wood together.

A coachman named James Barnett also saw the pair in a wood. 'They kissed each other several times and remained together about half an hour.'

Counsel for Mrs Tourle admitted that though it was impossible to deny that adultery had taken place, by taking back his wife in January, 1857, and living with her for eight months, Mr Tourle had condoned the offence. Lord Campbell however pointed out that even if condonation was proved for January, 1857, this could not apply to January, 1858.

After both counsels had made their final speeches, the jury announced they would not trouble his Lordship as they had already made up their minds. But Lord Campbell advised them they had better hear what he had to say about the questions they had to decide.

He then passed some comments on 'the law by which persons in all walks of life were now able to obtain a remedy that was formerly, on account of its expense, confined to the wealthy. . . . He made some strong remarks about the conduct of Mrs Tourle, pointing out that at the very time she was instituting proceedings in the court for restitution of conjugal rights, she was carrying on an adulterous intercourse with her paramour at the London-hotel. Mr Tourle, he added, appeared to be an honourable man, attached to his wife, and there appeared to be no good ground whatever for the charge that he had connived at his own dishonour.'

The jury found Mrs Tourle guilty of adultery and also that there had been no condonation, connivance or desertion by her husband. Costs were awarded against Mr Renshaw, the correspondent.

Young Renshaw who had purchased a cornetcy in the 2nd Dragoon Guards nine months before the court proceedings, exchanged soon after to the 13th Hussars. In June, 1859, he married his 'paramour' at St James Church, Paddington. The bride was now 30, but, exercising one of the few privileges then open to her sex, reduced it to 24 which was the age of her new husband.

Obviously the Renshaws were forced by the rigid moral conventions of their time to make a fresh start where their past was not known. But since both had shown an aptitude for intrigue, this may have been no great burden. In the spring of 1860 they joined the Inniskillings at Kirkee and for more than a year seem to have kept their past hidden.

Page 27   6. *Sir William Mansfield* (1818–76). Passed through Sandhurst. Appointed ensign in 53rd Foot (King's Shropshire Light Infantry) in 1835. Rapidly bought succeeding promotions. Lieutent-Colonel at 32 over the head of Henry Havelock, a hero of the Mutiny, 24 years his senior. Became C.-in-C. India in 1865 after Sir Hugh Rose. On elevation to the peerage assumed title of Lord Sandhurst and became Government Spokesman in the Lords for the abolition of the Purchase System.

Highly intelligent man but temperamentally unsuited to military command. His lack of temper and judgement made him unpopular. Had very defective eyesight, so was dependent on others for vital information in the field. Inability or unwillingness to trust those who gave it led to painful and discreditable quarrels with his subordinates.

It was said of him that he would have been a success in any profession but one—the Army.

Page 91   7. Colonel Crawley, having read out the finding, sentence and remarks of the Commander-in-Chief, then proceeded to supplement them by a speech of half an hour's duration. He opened his remarks by stating that 'seven months had elapsed since the beginning of the trial; that his honour and his character had been vilified; that he had been held up to public approbrium; that he had been through a fiery ordeal; that he had been accused of all the crimes of the Decalogue; that if he lived to the age of Methuselah, he would never forget it; that he took that favourable opportunity of contradicting all the false and malicious statements made against him.'

Then, with his hand raised aloft, and in a very loud voice, he said, 'that such a victory had never before been obtained, and that a righteous judgement had fallen on his enemies; that it had rarely happened to any commanding-officer to be eulogized as he had been by a Commander-in-Chief; that he had only been found fault with for being too lenient in his conduct towards officers when they first manifested a spirit of insubordination, but that he would take care not to make that mistake again; that he now had all those who dared to dispute his authority under his thumb; that if anyone dared to do so again, he would (this was said in his severest tone and with suitable gesticulation) crush him.' He then complained (all this on parade, in the presence of the men), that he had been ill-treated by his officers, and wound up his speech by thanking the men for their good conduct, and offered to lead them to glory.

(*Army and Navy Gazette*, 22 November, 1862.)

Page 100 8. Early in 1862, Lieutenant-Colonel Bentinck of the 4th Dragoon Guards, charged Captain Robertson, one of his officers, with 'conduct unbecoming an officer and a gentleman'. At the court-martial almost all the Regiment rallied to the defence of the Captain and evidence soon emerged to prove conclusively that Bentinck was himself guilty of the very conduct for which his subordinate was on trial—a turn of events which naturally delighted those who were demanding reforms in the Army.

Their joy was short-lived. Although he had not completed the necessary qualifying years of service, the Colonel was allowed to retire on half-pay with no charges brought against him and with no official stain on his character. The Captain was certainly acquitted. But two officers who gave evidence in his favour were actually punished by the Court: Major Jones had his promotion stopped and Lieutenant Rintoul was severely reprimanded 'for keeping a diary', the contents of which had been used to prove the prisoner's innocence.

There were indignant protests from outsiders but the verdict was allowed to stand.

Shute's being chosen to take over the Regiment indicates the Horse Guard's confidence in his ability to handle a delicate situation which in the event seems to have been justified.

Page 103 9. *James Matthew Higgins* (*1810–1868*). Social reformer. Rich. Educated Eton and New College, Oxford. Life was uneventful. Had no literary ambition and wrote only to redress a wrong, abolish an abuse or deflate pomposity. Practical and no sentimentalist. Went to help during the Irish famine. Got himself made a parish guardian to demonstrate what could be done.

Known as *Jacob Omnium* or J.O. Always wrote under a pseudonym to prove that prestige was not necessary to gain attention. Eg. Attacked Public Schools under 'Paterfamilias'; the Army under 'A Civilian'. His letter to *The Times* immediately fixed public attention on the Lilley case. His grasp of essential facts and ability to inject pungency into apparently simple statements made him the most formidable controversialist of his day on social issues. Close friend of Thackeray, who, as editor of *Cornhill Magazine* published some of his articles.

A 'gentle giant'. Height of six feet eight inches was frequent source of humour. Namesake, *only* six feet four, complained of being called 'Little Higgins'. Enjoyed good health until he became subject to attacks of rheumatism not long before he died unexpectedly after bathing in the sea in the summer of 1868.

Page 110 10. *Dudley Francis Fortescue* (*1821–1909*). Third son of 2nd Earl Fortescue. Liberal M.P. for Andover, 1857–74. Supported great reforms of Gladstone Administration. After Party's defeat in 1874 retired to his brother's estate in Ireland. Lifelong devotion to charitable work—Poor Law, Lunacy, N.S.P.C. Never sought limelight. Maintained interest in affairs and was receptive to new ideas up to his death at age of 88.

Page 113 11. *William Coningham* (*1815–1884*). Son of Rev. Robert Coningham of Londonderry. Born Penzance. Cornet in 1st Light Dragoons 1834–6. Independent Liberal M.P. for Brighton (1857–64). Advocate of radical reforms—the

ballot, extension of suffrage, *temperance*. Sincere and coura-
geous but excitable and ineffective as a speaker.

Page 158   12. *Naval and Military Intelligence.*

'. . . Apartments in the Club-house have been fitted up for
him (Colonel Crawley) to prevent the unpleasantness of his
removal to and from the court.

'On Friday, in accordance with the sentence of a court-
martial, the ceremony of "drumming out" a soldier was
performed at Aldershott. The culprit was Private O'Donnell
of the 76th Regiment, whose extraordinary attempt at
desertion was noticed in *The Times* about three months ago.
It will be remembered that O'Donnell escaped with his rifle,
bayonet and 20 rounds of ammunition in his possession into
the Long Valley, and for three days kept a pursuing party at
bay, being ultimately lost sight of in the direction of
Reading. Subsequently he quietly returned to his regiment,
was placed under arrest and tried by court-martial for the
offences. Being found guilty on the several counts of desertion, firing at a corporal of the 76th Regiment and a private
of the 6th Dragoon Guards, and of losing various articles of
regimental apparel, ammunition, etc, he was adjudged to
undergo "four years penal servitude, to be ignominiously
expelled her Majesty's service and to be branded with the
letter 'D' ''.

'Two companies of the 76th Regiment remaining at
Aldershott having assembled in the barrack square together
with the band of the 60th Rifles, the prisoner was marched
out hand-cuffed, wearing only his shirt and trousers, as when
last seen at the period of his escape. Since his arrest he has
constantly feigned insanity, and he now exhibited a very
defiant bearing. The companies being drawn into line and
the front rank marching fifteen paces to the front, Captain
O'Donoghue, the senior officer in command, read the charges
and the finding of the court-martial, together with the whole
list of offences recorded against the prisoner, from which it
appeared that his name had figured in the defaulter book 41
times and that he had previously at Glasgow, been sent to
112 days imprisonment and to be branded with the letter
"D".

'On the conclusion of the reading of the sentence, the
prisoner threw his cap high in the air and shouted "Long
live Queen Victoria! Good-bye! Three cheers for Old
Ireland."

'Quickly stepping out he was followed by the band of the 60th Regiment playing "The Rogues March". On arrival at the end of the line he turned round, bowed nearly to the ground, again shouted "Good-bye" and threw his cap into the air three or four times. He was then marched, under escort of the 60th Regiment, to the military prison to undergo the full term of punishment.'

(*The Times* 9 November, 1863.)

Page 158 13. *Sir William George Granville Venables Vernon Harcourt* (*1827–1904*). 'Vernon Harcourt'. Born Yorkshire. Educated Trinity College, Cambridge. Called to the Bar. Q.C. 1866. Liberal M.P. 1868. Solicitor-General under Gladstone, 1873. Home Secretary and Chancellor of the Exchequer—introduced graduated death-duties in 1894 Budget. Brilliant speaker. Famous remark—'We are all socialists now' (1892). Was expected to succeed Gladstone in 1894 but served on under Lord Rosebery. Retired from politics in 1898.

Page 158 14. *Sir Alfred Hastings Horsford* (*1818–85*). Rifle Brigade, 1839. Lieutenant-Colonel, 1853. Kaffir War (1852–3), the Crimea and in the Indian Mutiny commanded 3rd Battalion of the Rifle Brigade at Cawnpore and Lucknow.

Page 165 15. It did however evoke a letter to *The Times*.

Sir,

I beg to be forgiven for encroaching upon your valuable time as I only wish to direct your attention to a most surprising fact—namely, that during the long illness of Mrs Lilley and the death of her husband, Sergeant-major Lilley, which, according to Mrs Cotton's evidence was not at all sudden, and finally at her own death, the chaplain of the force seems never to have had anything to do with these persons.

Are gallant British soldiers and their families while stationed abroad in heathen countries allowed to be ill and to die without having some religious consolation offered to them? Besides, in Colonel Crawley's case, how very valuable the evidence of the chaplain might have been is needless to say.

I am, etc,

J. G. de S.

Page 177 16. *Detailed Medical Case of the late R.S.M. J. Lilley. Age 37 years. Service 18 years 4 months.*

A strong, healthy, well developed man, inclined to corpulency, about 17 or 18 stone weight. Had always enjoyed good health. Reported sick when in close arrest on the morning of the 24th May about 9 am, was immediately visited by Assistant-Surgeon Barnett at his quarters; he then complained of colicky pains and flatulent distension of abdomen, heaviness in his head, and feelings of oppression about the chest, attended with great depression of spirits. Bowels distended and rather confined; a large dose of castor oil was given and hot fomentation applied to the abdomen; this afforded temporary relief; during the day continued restless, but only referred his distress to griping pains and flatulent distension of abdomen. At 6 pm, the castor-oil not having acted sufficiently, and the abdominal distension and griping continuing, 8 grains of calomel was administered, and two hours after he had an anodyne and antispasmodic draught of hyoscyamus ether and camphor mixture.

Was seen again at 11.30 that night, when it was found that free evacuation from the bowels had not supervened since the evening visit. A large cathartic enema was then administered which afforded immediate relief to the abdominal uneasiness. Soon after this he became restless, moaned a little, eyes dull and heavy, with an anxious expression of countenance, cold affusion was applied to head and chest. When seen at 2.30 am he was insensible and breathing stertorously, pulse rapid and full, face flushed; the temporal artery was opened, and a few ounces of blood drawn. Cold affusion and spirit lotion continued to head and chest, and extremities immersed in hot water, and rubbed with mustard, &c., not the slightest reaction was induced. He died comatose about 4 am. (Remainder of Report quoted on page 94.)

Index